Sustainable Future

Editor: Danielle Lobban

Volume 439

independence
educational publishers

First published by Independence Educational Publishers

The Studio, High Green

Great Shelford

Cambridge CB22 5EG

England

© Independence 2024

ISBN-13: 978 1 86168 899 6

Printed in Great Britain

Zenith Print Group

Acknowledgements

The publisher is grateful for permission to reproduce the material in this book. While every care has been taken to trace and acknowledge copyright, the publisher tenders its apology for any accidental infringement or where copyright has proved untraceable. The publisher would be pleased to come to a suitable arrangement in any such case with the rightful owner.

The material reproduced in **issues** books is provided as an educational resource only. The views, opinions and information contained within reprinted material in **issues** books do not necessarily represent those of Independence Educational Publishers and its employees.

Images

Cover image courtesy of iStock. All other images courtesy of Freepik, Pixabay and Unsplash.

Additional acknowledgements

With thanks to the Independence team: Shelley Baldry, Tracy Biram, Klaudia Sommer and Jackie Staines.

Danielle Lobban

Cambridge, January 2024

Contents

Chapter 1: What Is Sustainability?

Chapter 2: The Circular Economy

Chapter 3: The Future?

Introduction

Sustainable Future is Volume 439 in the **issues** series. The aim of the series is to offer current, diverse information about important issues in our world, from a UK perspective.

About Sustainable Future

Sustainable practice is top of the agenda for global industries and small businesses. This book explores how we can live more sustainably, what the circular economy is, and how small changes can protect the world we live in for future generations.

Our sources

Titles in the **issues** series are designed to function as educational resource books, providing a balanced overview of a specific subject.

The information in our books is comprised of facts, articles and opinions from many different sources, including:

- Newspaper reports and opinion pieces
- Website factsheets
- Magazine and journal articles
- Statistics and surveys
- Government reports
- Literature from special interest groups.

A note on critical evaluation

Because the information reprinted here is from a number of different sources, readers should bear in mind the origin of the text and whether the source is likely to have a particular bias when presenting information (or when conducting their research). It is hoped that, as you read about the many aspects of the issues explored in this book, you will critically evaluate the information presented.

It is important that you decide whether you are being presented with facts or opinions. Does the writer give a biased or unbiased report? If an opinion is being expressed, do you agree with the writer? Is there potential bias to the 'facts' or statistics behind an article?

Activities

Throughout this book, you will find a selection of assignments and activities designed to help you engage with the articles you have been reading and to explore your own opinions. Some tasks will take longer than others and there is a mixture of design, writing and research-based activities that you can complete alone or in a group.

Further research

At the end of each article we have listed its source and a website that you can visit if you would like to conduct your own research. Please remember to critically evaluate any sources that you consult and consider whether the information you are viewing is accurate and unbiased.

Issues Online

The **issues** series of books is complemented by our online resource, issuesonline.co.uk

On the Issues Online website you will find a wealth of information, covering over 70 topics, to support the PSHE and RSE curriculum.

Why Issues Online?

Researching a topic? Issues Online is the best place to start for...

Librarians

Issues Online is an essential tool for librarians: feel confident you are signposting safe, reliable, user-friendly online resources to students and teaching staff alike. We provide multi-user concurrent access, so no waiting around for another student to finish with a resource. Issues Online also provides FREE downloadable posters for your shelf/wall/table displays.

Teachers

Issues Online is an ideal resource for lesson planning, inspiring lively debate in class and setting lessons and homework tasks.

Our accessible, engaging content helps deepen students' knowledge, promotes critical thinking and develops independent learning skills.

Issues Online saves precious preparation time. We wade through the wealth of material on the internet to filter the best quality, most relevant and up-to-date information you need to start exploring a topic.

Our carefully selected, balanced content presents an overview and insight into each topic from a variety of sources and viewpoints.

Students

Issues Online is designed to support your studies in a broad range of topics, particularly social issues relevant to young people today.

Thousands of articles, statistics and infographs instantly available to help you with research and assignments.

With 24/7 access using the powerful Algolia search system, you can find relevant information quickly, easily and safely anytime from your laptop, tablet or smartphone, in class or at home.

Visit issuesonline.co.uk to find out more!

What does it mean to live in a sustainable way?

The United Nations says that sustainability means 'meeting the needs of the present, without compromising the ability of future generations to meet their own needs'. In other words, people living today shouldn't use up the world's resources so that there are none left for those who live after us.

To sustain something simply means to keep it going. There are many people living on planet Earth - over 7.8 billion and counting! - and they all need certain things to survive. As a bare minimum, people need food, water and somewhere to live. All of these needs have to be met by the resources that are available on the planet. If we use up resources that can't be replaced, or we carry out practices that harm the planet faster than it can recover, this is not sustainable.

When we talk about living in a sustainable way we are generally talking about developing ways to live that don't harm the environment, or that cause the lowest possible impact on it. Living sustainably means that people try to manage their needs in a way that will allow future generations to do the same. Sustainable living means finding ways that we can make the planet's resources last as long as possible, ideally forever.

Some of the planet's resources are already running out or they are causing problems because there aren't enough for everyone who wants them. These are the problems that living sustainably tries to improve.

'Earth Overshoot Day' is the date each year when humanity has used up all the resources (such as forests or fish) that it would be possible for Earth to regenerate in that same year. Ideally, we would be looking to last a whole year without reaching that date. In fact, in 2020, Earth overshoot day was August 22nd. That means that every day of 2020 after that date, people were using more than the planet could sustain. This was actually an improvement on 2019's overshoot day (July 29th) as a result of the pandemic, which prevented human action, such as travel, for several weeks. It isn't possible to keep using resources like this forever. The world is currently using around one and three quarter Earths' worth of resources a year – but we don't have any more Earths!

2023

Brainstorm

- What do we mean by 'sustainability'?
- Why is achieving a sustainable lifestyle important?
- What is sustainable development?

Gen Z cares about sustainability more than anyone else – and is starting to make others feel the same

By Johnny Wood

- **Generation Z cares more about sustainable buying decisions than brand names, a new survey shows.**

- **This first generation of 'digital natives' is inspiring other age groups to act more sustainably.**

- **The main reasons for consumers not adopting a more eco-friendly lifestyle are a lack of interest because they think it's too costly or insufficient information.**

Generation Z shows the most concern for the planet's well-being and influences others to make sustainability-first buying decisions, according to new research.

GenZ are people born between the nineties and the noughties – roughly spanning 1995 to 2010 – and three-quarters of them prefer to buy sustainably rather than to go for brand names. That's according to respondents in a survey of US consumer attitudes on sustainable shopping by First Insight and the Baker Retailing Center at the Wharton School of the University of Pennsylvania.

Labelled the first 'digital natives' by McKinsey, but also known as the TikTok generation, most members of Gen Z are now in their mid-20s, are generally tech-savvy and are accustomed to making informed purchasing decisions.

As champions of sustainable consumer practices, Generation Z's views also influence other age groups to change their buying behaviour.

Across the generational divide, consumers are prepared to spend more on sustainable products now than two years ago, comparisons with data from an earlier survey show.

Spending on sustainable brands and products by Generation X – those born between the mid-1960s and mid-1970s – has increased by 24% since 2019. And the behaviour of other groups has followed a similar trend.

Gen Z's influence has led to it being labelled 'the most disruptive generation ever' by the Bank of America, a CNBC article reports.

Choosing a sustainable lifestyle

In our evermore connected world, consumer awareness is on the rise – including what, how and where we consume. And change is happening.

Saying no to single-use plastics is the most common sustainable lifestyle change among UK consumers, a 2021 survey by Deloitte shows. More than 60% have reduced their use of throwaway plastics.

How do you rate the importance of these factors when making a purchase?

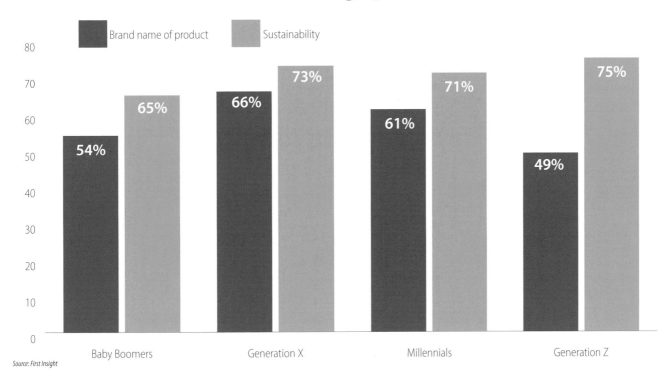

Legend: Brand name of product / Sustainability

Baby Boomers: 54% / 65%
Generation X: 66% / 73%
Millennials: 61% / 71%
Generation Z: 49% / 75%

Source: First Insight

Almost two-fifths of those surveyed have reduced the number of flights they take, and the same proportion said they are buying fewer new goods.

Obstacles to sustainability

So what's preventing more people from switching to more sustainable lifestyles?

The main reason for not embracing positive change – given by two-fifths of UK respondents – is a lack of interest, followed by a segment of 16% who said it was too costly, and a further 15% who want more information before considering lifestyle changes.

As the climate crisis continues to impact our lives, awareness of the need to act sustainably is growing among consumers of all ages, albeit at different rates. Brands and retailers who respond to consumer demand to think and act more sustainably will be better placed to succeed in an increasingly uncertain future.

18 March 2022

Prefer to buy from sustainable brands

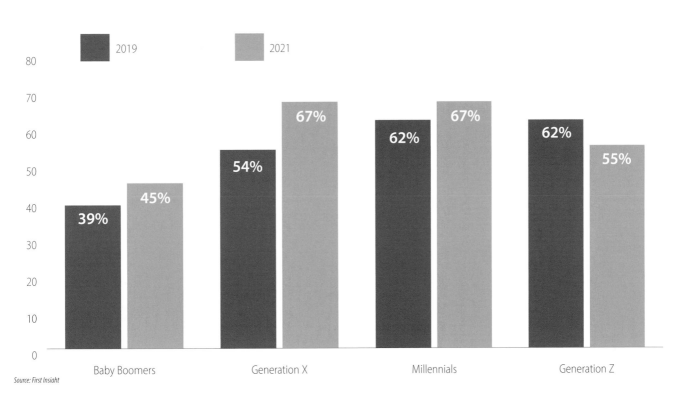

Legend: 2019 / 2021

Baby Boomers: 39% / 45%
Generation X: 54% / 67%
Millennials: 62% / 67%
Generation Z: 62% / 55%

Source: First Insight

In the last 12 months which of the following have you personally done, specifically in an effort to adopt amore sustainable lifestyle?

% of all UK consumers

Yes, I have to adopt a more sustainable lifestyle
No, I haven't
Don't know/can't recall
Not applicable - I have done this in the past 12 months, but not to adopt a more sustainable lifestyle

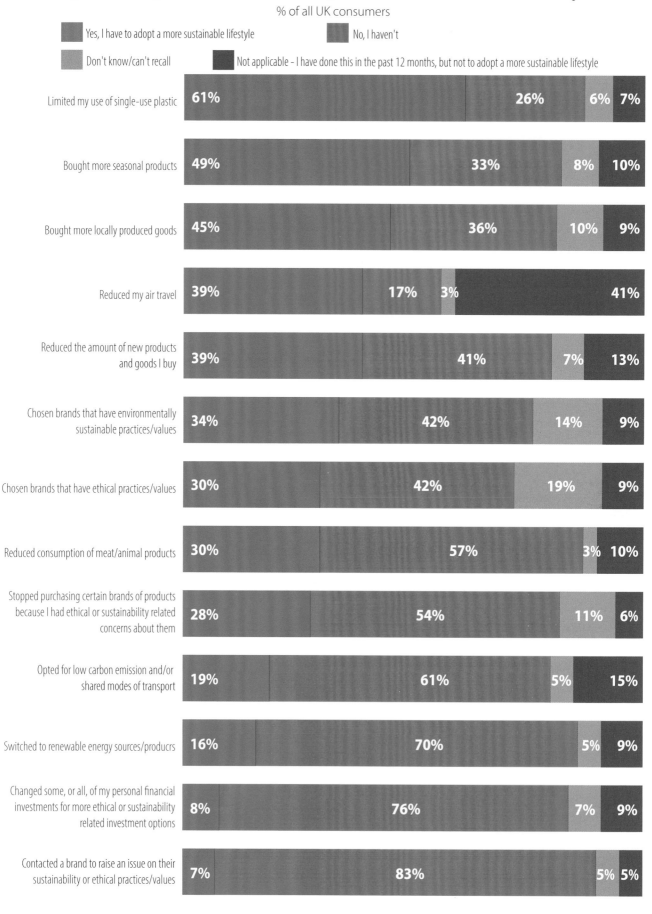

	Yes	No	Don't know	Not applicable
Limited my use of single-use plastic	61%	26%	6%	7%
Bought more seasonal products	49%	33%	8%	10%
Bought more locally produced goods	45%	36%	10%	9%
Reduced my air travel	39%	17%	3%	41%
Reduced the amount of new products and goods I buy	39%	41%	7%	13%
Chosen brands that have environmentally sustainable practices/values	34%	42%	14%	9%
Chosen brands that have ethical practices/values	30%	42%	19%	9%
Reduced consumption of meat/animal products	30%	57%	3%	10%
Stopped purchasing certain brands of products because I had ethical or sustainability related concerns about them	28%	54%	11%	6%
Opted for low carbon emission and/or shared modes of transport	19%	61%	5%	15%
Switched to renewable energy sources/producrs	16%	70%	5%	9%
Changed some, or all, of my personal financial investments for more ethical or sustainability related investment options	8%	76%	7%	9%
Contacted a brand to raise an issue on their sustainability or ethical practices/values	7%	83%	5%	5%

The above information is reprinted with kind permission from WORLD ECONOMIC FORUM
© 2023 World Economic Forum

Source: Deloitte

www.weforum.org

Five ways the new sustainability and climate change strategy for schools in England doesn't match up to what young people actually want

An article from The Conversation.

By Elizabeth Rushton, Associate Professor of Education, UCL and Lynda Dunlop, Senior Lecturer in Science Education, University of York

The UK government has introduced a new sustainability and climate change strategy for schools. However, our research shows that it does not go far enough to meet what young people and teachers want.

Last year, together with colleagues, we conducted research with over 200 teachers, teacher educators (the people who train teachers) and young people aged 16-18 from the UK to understand how they wanted schools to tackle sustainability and climate change. Participants were recruited via email and Twitter.

Our research allows us to assess how far the government's new strategy aligns with what teachers and young people want. Here are five key things that teachers, teacher educators and young people would like to see in schools – and how the government's sustainability and climate change strategy matches up.

1. Sustainability education for all

Many teachers already provide opportunities for pupils to learn about sustainability, such as eco-clubs, recycling projects and sustainable fashion shows. However, this work is optional and tends to happen outside the curriculum, meaning that not all young people have opportunities to take part.

Teachers and young people in our research wanted environmental sustainability to feature across the curriculum, not just in geography (which not all students study after the age of 14) and science.

The government's strategy includes a new natural history GCSE, which will be taught from 2025. This will increase opportunities for young people to learn about the natural world and sustainability. However, this subject will be optional and so will not ensure that every young person has access to climate change and sustainability education, regardless of their age or subject choice.

The government's new strategy does include other ways to learn about the environment. Pupils can take part in a climate leaders award, carrying out extra-curricular activities in connection with sustainability, but this is also optional. This means that environmental sustainability remains unlikely to be prioritised or to involve everyone.

2. Training for teachers

Teachers we spoke to wanted professional development opportunities so they could feel more confident teaching sustainability in the classroom. As one teacher said: 'We can lack confidence because we are navigating this ourselves and do not feel like experts where we might in our subject.'

While the new strategy offers support for teachers through resources and training, there is no promise of time to access this, and there is no fundamental change to existing school or teacher education curriculums in England.

3. Put knowledge into action

Teachers and young people do not just want to pass on knowledge – they want to be able to make a difference. We

heard that teachers and students wanted education to be more about critical thinking, data literacy, doing research, taking action and communicating and networking with others. As one young person said:

'We should be taught about big business and corporations – what their impact actually is. A lot of greenwashing goes on with big companies making individuals feel as if they are solely responsible... Education should empower us to demand change and to demand the rights we should have.'

The focus of the government's strategy is on learning more about sustainability, climate change and the natural world, not empowering young people to act for the environment or challenging the root causes of climate change.

4. Make schools sustainable

Teachers and young people wanted greater attention to environmental sustainability in school operations, including handling of energy, waste, transport and food. There is currently little requirement for schools in England to learn about or act for environmental sustainability.

The government's strategy focuses on net zero targets and promises action on waste by requiring schools to increase recycling and reduce landfill. It also promises at least four new low-carbon schools and one college.

In other aspects of school operations – food, transport and energy - there is encouragement and support in the strategy, which may or may not translate into action.

5. Make schools community hubs for climate action

Young people and teachers saw schools as community hubs where people from across different generations could take part in sustainability focused activities. They saw starting sustainability education with young children and incorporating this throughout their lives as vital.

Introducing the climate leaders award provides a way for the contribution young people are already making to environmental sustainability in schools and communities to be recognised and valued. The young people we worked with called for such a scheme and wanted it at no cost. However, the description of the climate leaders award in the government's sustainability and climate change strategy references existing awards such as the Duke of Edinburgh's award, which is not free of charge.

Teachers and young people told us that at present, there is little support for environmental sustainability in education. The government's new strategy does little to change this status quo.

We need further change to put sustainability and climate change at the heart of education. This could be done by putting climate change and sustainability into the core curriculum, making it part of exam specifications and school inspections and part of the core framework for teacher training – in other words, the things that teachers must prioritise.

4 May 2023

10 social sustainability examples (and why they matter)

Social sustainability examples are all around us and learning to recognize them and why they matter is crucial for understanding social sustainability. What is social sustainability?

Social sustainability is one of the 3 pillars of sustainability (also known as the 3 Ps) and it describes actions that we can take to improve society. Social sustainability issues examples include improving the quality of our lives, reducing inequality, cultural sustainability, and helping people make better choices about their health, education, and work.

Social sustainability examples

What are some examples of social sustainability?

Here are the 10 best social sustainability examples in our society:

1. Childcare
2. Social equity
3. Education
4. Inclusion
5. Community Outreach
6. Poverty alleviation
7. Senior care
8. Healthcare
9. Sport
10. Diversity in the workplace

Now, let's learn more about each of those social sustainable development examples and why they matter so much for a more sustainable society.

1. Childcare

Childcare is an issue that affects everyone. You may not have children, but it's still a challenge for your friends and family members who do. It's also a challenge for society as a whole.

Childcare is a universal challenge because it impacts the economy, the environment, and the community at large.

That's why childcare is an important example of social sustainability. Thanks to childcare services it is possible to help single parents to keep their jobs and income as well as provide new opportunities to socialize.

The challenge is particularly acute in the United States, where childcare is expensive and access to it is unequal. This can lead to consequences such as parents feeling pressure to leave their jobs or children being left at home unsupervised.

Childcare also has an impact on the economy: The amount of time that working parents spend with their children has a significant economic effect on them and on society as a whole. For example, one study found that the introduction of parental leave policies for working mothers had a 'positive impact on the working hours during the first 3 years after childbirth'.

Finally, a recent study shows that applying the UN sustainable development goals (SDG) framework to childcare can help to highlight the right steps to take when implementing childcare for sustainable development.

2. Social equity

Social equity is about everyone having equal access to the same opportunities. In other words, social equity is about everyone being able to participate in society and have a voice in the decision making that affects their lives.

Social equity is not just about treating people equally or equally well: it's also about treating them with respect and fairness.

Social sustainability includes tangible goals such as reducing poverty and increasing education levels, but it also includes intangible aspects like how we feel when we interact with our communities because isolation can be as harmful as hunger!

This means that social sustainability encompasses things like community empowerment and financial inclusion (the ability for all people regardless of income level to participate in society), but it can also mean having fun through shared experiences like playing sports together or attending cultural events together (which builds trust between neighbors).

For example, services to ensure social justice like the ones offered by the American Civil Liberties Union Foundation (ACLU) are a wonderful example of social sustainability in the United States.

3. Education

Education provides individuals with opportunities for personal growth and development, as such, it is one of the greatest social sustainability examples.

Social sustainability is an important part of education because it helps people to learn about their community and environment, which can help them to be more environmentally conscious.

Education is important for social sustainability because it also helps people to learn new skills, so they can be more productive in their jobs or other activities.

Additionally, schools, colleges, and universities are also important places where long-lasting social relationships are created for the new generations. Contributing to more cohesive communities and improving social sustainability.

4. Inclusion

Inclusion is about making sure that everyone has access to the same opportunities, and is treated with respect and dignity.

When we talk about inclusion, we're talking about how people are included in society. Inclusive societies are more likely to be successful because they include people from diverse backgrounds and make sure everyone has the same rights.

Let's look at some practical examples of social sustainability leveraging inclusivity:

- Providing transportation for families who can't afford a car so they can go grocery shopping or get to work

- Offering free tutoring programs for children from low-income families so they can do better in school

- Giving homeless people food, shelter, healthcare services

The best way to make sure everyone has access to the same things is to include people from different backgrounds and make sure they have equal rights.

The goal is for everybody to feel included in society, regardless of their social status or economic background.

5. Community Outreach

Community outreach is a way to improve the community by helping people. It can be as simple as giving out food, or it can be more complex, like providing education and training.

The goal of community outreach is always to improve society, and therefore social sustainability.

For example, a very popular organization that provides improvements of social sustainability through community outreach is Habitat for Humanity. Providing microfinance, shelter, and affordable housing in the United States, Canada, Europe as well as in developing countries.

6. Poverty alleviation

Poverty alleviation is a broad term for the reduction of poverty and economic inequality. It is about improving the

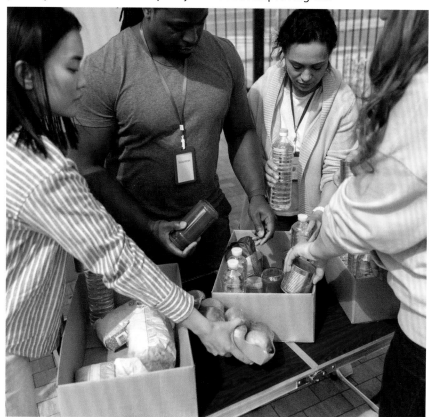

lives of people who are poor, allowing them to have more opportunities, creating a fairer society, and improving their wellbeing.

Poverty eradication is also one of the UN sustainable development goals, confirming its importance for social sustainability and a better world.

There are many ways you can help alleviate poverty and create a fairer society. You could start by taking up a voluntary role at your local charity or community center.

This will allow you to meet people who are poor and learn more about their needs. It is also a way of giving back to the community and creating new opportunities for those in need.

NGOs such as The Organization for Poverty Alleviation and Development (OPAD) are offering an important example of social sustainability around the globe. For example, one of their recent projects was to help prevent poverty due to drought in Zambia, which benefited about 20'000 people.

Another interesting perspective is also offered by this study, which shows how social sustainability can be improved in supply chains by implementing supplier development programs through NGOs.

7. Senior care

One example of social sustainability is senior care. Senior care is about creating communities where seniors can live their lives to the fullest, age with dignity and respect, and stay engaged in the community.

Many seniors are not able to leave their homes or need assistance with daily tasks such as cooking, cleaning, or laundry. In these cases, they may want to move into a facility that provides this kind of service.

This is a pressing issue, especially in the US and other developed countries, which have an aging population that will require more and more senior care services in the coming decade.

8. Healthcare

Healthcare is a great example of social sustainability because healthcare is a basic human right and should be available to everyone, regardless of income.

It's an example of a public good, which means that it benefits the entire population and not just those who pay for it.

Healthcare is also an example of public service because it involves providing people with access to certain goods or services that they would otherwise have difficulty purchasing on their own.

Lastly, this service creates social capital by helping individuals create strong relationships with one another; this type of relationship may lead them to feel more connected within their communities

A recent study also explored the use of a rating system to help improve the social

aspects of healthcare. This looks very promising because thanks to its use they were able to collect organizational and design suggestions to improve healing and the social sustainability of the facilities.

9. Sport

In the context of sport and social sustainability, a number of strategies can be used to achieve social sustainability by leveraging the social aspects of sports activities. These include:

- Sport can be used to promote social inclusion by encouraging participation among all sections of society, including those most disadvantaged. This is called horizontal integration.

- Sport promotes vertical integration through elite athlete development programs aimed at identifying talent and developing it for international competition. This ensures that high-performance athletes are representative of their country's populations in terms of gender, ethnicity, age, and other characteristics (such as disability). As well as fostering pride in one's country at an international level. Vertical integration encourages players from lower levels within sports organizations (eg junior players) to aspire towards reaching the highest level possible through their own excellence on the field or court.

In addition to these two forms of integration within sports organizations themselves (horizontal and vertical), sports can contribute directly towards promoting social justice outside them as well.

Moreover, sport is one of the best examples of social sustainability, because it encourages the creation of social relationships, can promote diversity, inclusion, equality, and provide a way out of poverty for athletes.

10. Diversity in the workplace

Here are some examples of social sustainability in the workplace!

Diversity and social sustainability. In order to be socially sustainable, a community must include a wide variety of people with different ideas, cultures and backgrounds.

Diversity is the cornerstone of social sustainability because it leads to more innovation, creativity, and better decisions.

A great example of this is diversity in the workplace. When employers hire a variety of individuals with different knowledge bases, they can often come up with better solutions than they could if they were all working in isolation from each other or had less experience in their field than others.

This is why many companies have become more intentional about hiring employees who represent different demographics from one another. So that everyone has access to ideas from diverse perspectives rather than having one person speak for everyone else at work (or worse yet not talking at all).

So much so that diversity in the workplace is also one of the key parameters considered by ESG metrics (Environmental, Social, and Governance).

Including women and people with diverse backgrounds on the board of directors is an important aspect of Corporate Social Responsibility (CSR) and the social sustainability of an organization. This is because a more diverse top management will be able to take better decisions.

Conclusion

Social sustainability is important because it helps us to live, work and play in a just and fair society. Hopefully those social pillar of sustainability examples helped you to better understand this fundamental principle.

Together with human, economic, and environmental sustainability, social sustainability is one of the 4 pillars of sustainability. That's why achieving social sustainability and knowing great examples of how to implement it is so important: without it, we would not be able to get sustainable development!

We can use the social sustainability examples above to think about how we might apply these ideas in our own communities, workplaces, and businesses to make a better world and society.

19 August 2022

UK considers using new international rules to crack down on greenwashing

The International Sustainability Standards Board published its inaugural standards on Monday to help countries regulate companies' green claims.

By Rebecca Spree-Cole

The UK is considering adopting new rules aimed at clamping down on corporate greenwashing.

The International Sustainability Standards Board (ISSB) – a group set up at Cop27 to set global rules on climate reporting – published its inaugural standards on Monday to help countries regulate companies' green claims.

Under the rules, firms would face more pressure to publicly disclose their impact on the climate including on their Scope 3 emissions – which covers the products or services that they sell.

The ISSB says the aim is to establish a common global language in which companies report their impact, to improve trust in climate reporting and to help to inform investors about sustainability-related risks and opportunities.

> **'We have been really encouraged by the number of jurisdictions that have already indicated they will consider adoption.'**
>
> – Emmanuel Faber, ISSB chair

Speaking at the launch event on Monday, ISSB chair Emmanuel Faber said individual countries can decide whether listed companies must apply the new rules.

The UK is among the countries considering their use, alongside Canada, Japan, Singapore, Nigeria, Chile, Malaysia, Brazil, Egypt, Kenya and South Africa, according to the ISSB.

The London Stock Exchange and the Financial Conduct Authority (FCA) – the UK's financial watchdog – have both welcomed the new standards with the latter having worked closely with the ISSB on the new rules.

Mr Faber told the PA news agency: 'We have been really encouraged by the number of jurisdictions that have already indicated they will consider adoption.

'The UK has been instrumental in paving the way for sustainability reporting for investors and advising on the development of ISSB standards.

'We launched our standards today at the market opening of the London Stock Exchange, as well as at other exchanges around the world, and we have been in close dialogue with the UK FCA which is highly supportive of our objective to establish a common language for investors globally.'

The ISSB is part of the independent International Financial Reporting Standards Foundation, which writes standardised accounting rules used in around 140 countries.

> ### 'We encourage policymakers to adopt the ISSB's new standards as this global baseline by 2025.'
>
> – Jane Goodland, London Stock Exchange Group

It says the new rules on sustainability-related reporting can be released alongside firms' annual financial reports from 2024.

Sacha Sadan, director of ESG at the Financial Conduct Authority, said: 'We have been working closely with the ISSB since the start and are hugely supportive of its mission to create a common, global language for companies around the world to communicate their sustainability stories in a consistent and comparable way.

'That is why we are delighted to see the final standards launched today.'

Jane Goodland, group head of sustainability at the London Stock Exchange Group (LSEG), said: 'Corporate disclosure of basic metrics such as carbon emissions vary significantly based on size, sector and location and has not improved materially in recent years.

'This is impeding those investors wishing to allocate capital based on sustainable investment objectives.

'We encourage policymakers to adopt the ISSB's new standards as this global baseline by 2025.'

> ### 'It is important that investors, jurisdictions, listing authorities, companies and others consider how to incentivise adoption so there is a level playing field of comparable information.'
>
> – Gilly Lord, PwC

Pankaj Bhatia, director of Greenhouse Gas Protocol at the World Resources Institute, said: 'The ISSB's requirement to disclose Scope 3 emissions is a major step forward in measuring and managing emissions from companies' value chain.

'This is the first time a major global standard setting institution required reporting of Scope 3 emissions, setting a precedent for other institutions and regulatory programmes to follow.'

Financial services giant PwC welcomed the new rules but added that high quality standards alone are not enough.

Gilly Lord, global leader for public policy and regulation at PwC, said: 'Companies need to use them to produce high-quality sustainability reports.

'It is important that investors, jurisdictions, listing authorities, companies and others consider how to incentivise adoption so there is a level playing field of comparable information.'

She added that climate change is not the only area where investors and stakeholders need more reliable, comparable disclosures of non-financial information.

'So we encourage the ISSB to pursue its plan to go beyond climate and prepare standards with specific disclosure requirements focused on other sustainability areas,' she added.

26 June 2023

Brainstorm

In small groups, can you think of any examples of greenwashing? Do you know of any brands that sell 'green' products but are using the label to make the products more attractive to consumers?

What are the sustainable development goals?

'This is the people's agenda, a plan
of action for ending poverty in all its
dimensions, irreversibly, everywhere,
and leaving no one behind.'

– Ban Ki-moon

Here are the facts:

- The Sustainable Development Goals (SDGs) are the set of 17 agreed goals which all 193 UN member states including the UK committed to that will guide policy and funding for the next 10 years.

- According to the World Bank, around 1 billion people still live in extreme poverty and more than 800 million people do not have enough food to eat.

- From 2016, all UN member states committed to mobilise efforts to end all forms of poverty, fight inequalities and tackle climate change, while ensuring that no one is left behind.

The 17 sustainable development goals

Goal 1: End poverty in all its forms everywhere

Goal 2: End hunger, achieve food security and improved nutrition and promote sustainable agriculture

Goal 3: Ensure healthy lives and promote well-being for all at all ages

Goal 4: Ensure inclusive and equitable quality education and promote lifelong learning opportunities for all

Goal 5: Achieve gender equality and empower all women and girls

Goal 6: Ensure availability and sustainable management of water and sanitation for all

Goal 7: Ensure access to affordable, reliable, sustainable and modern energy for all

Goal 8: Promote sustained, inclusive and sustainable economic growth, full and productive employment and decent work for all

Goal 9: Build resilient infrastructure, promote inclusive and sustainable industrialization and foster innovation

Goal 10: Reduce inequality within and among countries

Goal 11: Make cities and human settlements inclusive, safe, resilient and sustainable

Goal 12: Ensure sustainable consumption and production patterns

Goal 13: Take urgent action to combat climate change and its impacts

Goal 14: Conserve and sustainably use the oceans, seas and marine resources for sustainable development

Goal 15: Protect, restore and promote sustainable use of terrestrial ecosystems, sustainably manage forests, combat desertification, and halt and reverse land degradation and halt biodiversity loss

Goal 16: Promote peaceful and inclusive societies for sustainable development, provide access to justice for all and build effective, accountable and inclusive institutions at all levels

Goal 17: Strengthen the means of implementation and revitalize the global partnership for sustainable development.

The content of this publication has not been approved by the United Nations and does not reflect the views of the United Nations or its officials or Member States.

SUSTAINABLE DEVELOPMENT GOALS

1 NO POVERTY

2 ZERO HUNGER

3 GOOD HEALTH AND WELL-BEING

4 QUALITY EDUCATION

5 GENDER EQUALITY

6 CLEAN WATER AND SANITATION

7 AFFORDABLE AND CLEAN ENERGY

8 DECENT WORK AND ECONOMIC GROWTH

9 INDUSTRY, INNOVATION AND INFRASTRUCTURE

10 REDUCED INEQUALITIES

11 SUSTAINABLE CITIES AND COMMUNITIES

12 RESPONSIBLE CONSUMPTION AND PRODUCTION

13 CLIMATE ACTION

14 LIFE BELOW WATER

15 LIFE ON LAND

16 PEACE, JUSTICE AND STRONG INSTITUTIONS

17 PARTNERSHIPS FOR THE GOALS

Source: United Nations

www.un.org/sustainabledevelopment

UK fails to progress on SDGs

A newly released report has revealed that, since 2018, the UK has only made good progress on 17% of its targets under the UN Sustainable Development Goals (SDGs).

By Heather Dinwoodie

* **The UK has only made progress towards 23 of its SDG targets, while stalling on 65 and regressing on 18.**

* **Its findings come as the government is introducing a range of policy measures that risk taking the country even further off track.**

* **The achievement of the SDGs, by the UK and other nations around the world, will demand a far more holistic perspective that considers specific contexts, stakeholder influence and the various interconnections between the goals and their associated targets.**

The UN Global Compact Network UK has released its *Measuring Up* report, which comes as the first complete update to its SDG progress series since 2018. The report assesses the country's performance against each of the 17 goals and their collective 169 targets.

It concludes that, overall, the UK's progress towards the SDGs has been fairly stagnant over the four-year period. Although the country has improved on 23 targets including the increase of renewable energy, tightening of financial legislation and reduction of food waste, it has failed to make notable progress towards 65 targets which were rated as 'red' or 'amber' under the traffic light system of 2018's report.

No change was found in the integration of climate considerations under national policy, nor on raising awareness around climate-related issues. Progress has also stagnated on marine pollution and marine ecosystem protection, the promotion of sustainable forest and land management and the overarching sustainable management of natural resources.

Furthermore, previous progress towards 18 of the targets has now regressed. Specifically, the report notes that the number of people living in poverty is on the rise.

It's bad, but will it get worse?

These findings are alarming enough at first glance, but they become all the more significant when considering the UK's current political context. Far from introducing measures that could accelerate progress towards the SDGs, the government's recent policy decisions have sparked major concerns.

In a bid to address soaring energy prices, Liz Truss's conservative government plans to accelerate North Sea oil and gas exploration, allow fossil fuels companies to continue making record breaking profits rather than be subjected to a windfall tax, resume the country's fracking activity and

amend or remove hundreds of environmental protection laws inherited from the EU.

The government's decisions directly contradict the demands of environmental organisations and the private sector, with Greenpeace preparing for legal action while the Corporate Leaders Group (CLG) has called for economic resilience to be built through an integrated approach that prioritises decarbonisation and the restoration of nature.

Its members recognise the positive impact that such an approach could have not only for the environment, but also for the UK's social and economic resilience.

The UK in comparison to other countries

To put the UK's lack of progress towards the SDGs into context, it helps to consider the framework's success in other countries. The SDGs were designed to be globally inclusive, enabling them to be integrated with the national development strategies of the 193 nations that pledged to adopt them.

This has proved problematic, however, as different countries face different challenges and, in most, the 15-year timeframe spans a number of changes in government. The universal nature of the SDGs means that national governments must attempt to define their development plans using a single set of metrics, which may not account for their specific needs.

When considering the agricultural sector, for example, the SDGs offer broad measures for productivity and resource access but do not provide metrics on individual crops or technological developments that might be relevant to particular areas.

Furthermore, unforeseen crises such as the COVID-19 pandemic and various outbreaks of crisis have impacted countries around the world, with interlinked consequences on food and nutrition, health, education, climate action and global peace.

These crises, compounded by the cascading risks of climate change, have interrupted progress towards the SDGs throughout the world. The UN's international 2022 report details how years of improvements in poverty levels, hunger, health, education and basic services have been reversed, and calls for urgent action to rescue both the SDGs and the 1.5C° limit set by the Paris Agreement.

With countries seemingly failing to deliver the SDGs, the private sector has made various attempts to align the framework with corporate targets. Asset management firm Robeco has said that measuring companies' contributions towards the SDGs will allow investors to assess the long-term, positive impacts of their decisions, while data analytics start-up Util has attempted to quantify the positive and negative SDG contributions of different investment funds.

Although the actions of the private sector do affect global progress towards the SDGs, it is important to remember that the framework was designed for application by national governments. There is rarely a simple answer as to whether companies contribute positively or negatively, as each industry has a range of operational impacts and wider outcomes.

Where next for the SDGs?

While the report's findings raise questions as to whether national targets could ever truly deliver sustainable development, the SDG framework remains useful as a means of tracking and improving on progress. Robust, consistent and measurable targets are undoubtedly necessary in determining where countries want to go, and how they will get there.

The main challenge is the development of strategies that acknowledge the interconnections between each goal. The transformational vision of the SDGs relies on such a holistic perspective, which identifies possible synergies while addressing potential trade-offs between targets.

Social, economic and environmental systems are deeply interconnected, with the UN itself acknowledging that, 'the 2030 Agenda for Sustainable Development states that the Sustainable Development Goals (SDGs) are integrated and indivisible and balance the three dimensions of sustainable development (economic, social, and environmental).

'The interlinkages and integrated nature of the SDGs are of crucial importance in ensuring that the purpose of the new Agenda is realised'. The UK report supports this, concluding with 120 individual recommendations on how the UK can advance its progress towards the SDGs. It identifies holistic planning, strong cross-departmental leadership and stakeholder engagement as three core priorities.

'The SDGs have tremendous potential to mobilise action across the whole of society but both government and business are missing an opportunity to use the holistic framing of the SDGs to address systemic challenges', it says, 'we see clear evidence that the strong interconnections between the Goals (and their Targets) mean that tackling these issues will require systemic change. The Goals cannot be achieved in isolation.'

Such recommendations raise hope that the UK, and other nations around the world, could improve their progress towards the delivery of the SDGs.

Doing so, however, will demand a holistic overview of the specific contexts of different locations, the contributions and limitations of stakeholders such as the private sector, and the implementation of national strategies that recognise how each goal interlinks with the others.

3 October 2022

The US committed to meet the UN's Sustainable Development Goals, but like other countries, it's struggling to make progress

An article from The Conversation.

By Scott Schang, Director of Environmental Law and Policy Clinic; Professor of Practice, Wake Forest University and John Dernbach, Professor of Law Emeritus, Widener University

I n a Zen parable, a man sees a horse and rider galloping by. The man asks the rider where he's going, and the rider responds, 'I don't know. Ask the horse!'

It is easy to feel out of control and helpless in the face of the many problems Americans are now experiencing – unaffordable health care, poverty and climate change, to name a few. These problems are made harder by the ways in which people, including elected representatives, often talk past each other.

Most people want a strong economy, social well-being and a healthy environment. These goals are interdependent: A strong economy isn't possible without a society peaceful enough to support investment and well-functioning markets, or without water and air clean enough to support life and productivity. This understanding – that economic, social and environmental well-being are intertwined – is the premise of sustainable development.

In 2015, the United Nations General Assembly unanimously adopted 17 Sustainable Development Goals, known as the SDGs, with 169 measurable targets to be achieved by 2030. Though not legally binding, all nations, including the U.S., agreed to pursue this agenda.

The world is now halfway to that 2030 deadline. Countries have made some progress, such as reducing extreme poverty and child mortality, though the COVID-19 pandemic set back progress on many targets.

On Sept. 18-19, 2023, countries are reviewing global progress toward those goals during a meeting at the United Nations. It's a good opportunity for Americans to review their own progress because, as we see it, sustainable development is fundamentally American.

Environment, economy and health intertwined

Though not widely recognized, sustainable development has been a core American policy since President Richard Nixon signed the National Environmental Policy Act into law in 1970. The law says that Americans should 'use all practicable means and measures … to create and maintain conditions under which man [sic] and nature can exist in productive harmony and fulfill the social, economic, and other requirements of present and future generations of Americans.'

While it is tempting in today's sour political climate to dismiss this as wishful thinking, the U.S. has made some progress reconciling economic development with environmental protection.

Gross domestic product, for example, grew 196% between 1980 and 2022, while total emissions of the six most common non-greenhouse air pollutants, including lead and sulfur dioxide, fell 73%, according to the Environmental Protection Agency.

The 2022 Inflation Reduction Act, a major sustainable development law, is designed to further accelerate the use of renewable energy and reduce greenhouse gas emissions through tax credits and other incentives. Goldman Sachs

estimated the law would spur about US$3 trillion in renewable energy investment. The law has already been credited with creating 170,000 new jobs and leading to more than 270 new or expanded clean energy projects. That impact further demonstrates that environmental goals can align with economic growth.

The 2015 Sustainable Development Goals cover a broader range of environmental, social and economic issues, and there are indicators for assessing progress on each.

How is America doing?

The U.S. ranked 39th out of 166 countries in a 2023 review of national efforts to implement the Sustainable Development Goals.

The Sustainable Development Solutions Network, which operates under the auspices of the U.N. Secretary-General, finds that America is lagging behind the targets set for many of the Sustainable Development Goals that are critical to the nation's defense, competitiveness and health, such as reducing obesity, increasing life expectancy at birth, protecting labor rights, reducing maternal mortality, decreasing inequality and protecting biodiversity.

To understand where the U.S. is falling short, we asked 26 experts working on various areas of sustainable development to review the nation's progress and make recommendations for future action. The resulting 2023 book, Governing for Sustainability, provides some 500 U.S.-specific recommendations for achieving the Sustainable Development Goals.

Health and access to quality health care loom large in many of the goals. The authors in several chapters explain why the nation cannot eliminate poverty or hunger, or have a vibrant economy, gender equality or education gains, without widely available, affordable health care. Yet, the U.S. has some of the highest health care costs in the world. Several states have rejected efforts to expand eligibility for federal Medicaid health insurance for low-income residents, leaving many people without care.

Similarly, the authors show that human health, ecological health, clean water and economic vitality all require sound climate policy. A quickly warming world poses new health risks, decimates ecosystems, strains potable water supplies and reduces global economic productivity.

Clean and abundant water is critical to a functioning economy and a stable, diverse ecosystem, and yet some areas of the United States still lack clean water or indoor plumbing. This often occurs in communities of color and low income, and it can impede economic prosperity and development in these areas.

Ready access to nutritious food is also a bedrock need to support many of the Sustainable Development Goals, from poverty alleviation to education, yet far too many American children rely on school lunches for basic sustenance.

The goals covering peace, justice, strong institutions and partnerships are necessary to achieve all of the goals. A society at war with itself and without rule of law cannot support a vibrant, diverse economy and lasting democracy. This has been shown repeatedly as some developing nations backslide from democratic progress and prosperity to civil war and poverty. Developed nations are subject to the same forces.

Taking the reins

Sustainable development is emphatically not about government alone solving the nation's problems. Businesses, universities and other organizations, as well as individuals, are essential to help the country realize its environmental, health and climate goals, fair practices and living wages.

The right place to 'take the reins' is where you are, and with the problems or tasks in front of you – at work and at home. Figure out more sustainable ways to use water and energy, for example. Look at what our book recommends and what others are already doing to meet the Sustainable Development Goals. Seize opportunities such as saving money, and reduce risks by, for example, cutting greenhouse gas emissions. Every individual can contribute to a better future.

6 September 2023

THE CONVERSATION

Circular Economy

Chapter 2

Circular economy: definition, importance and benefits

The circular economy: find out what it means, how it benefits you, the environment and our economy.

The European Union produces more than 2.2 billion tonnes of waste every year. It is currently updating its legislation on waste management to promote a shift to a more sustainable model known as the circular economy.

But what exactly does the circular economy mean? And what would be the benefits?

What is the circular economy?

The circular economy is a model of production and consumption, which involves sharing, leasing, reusing, repairing, refurbishing and recycling existing materials and products as long as possible. In this way, the life cycle of products is extended.

In practice, it implies reducing waste to a minimum. When a product reaches the end of its life, its materials are kept within the economy wherever possible thanks to recycling. These can be productively used again and again, thereby creating further value.

This is a departure from the traditional, linear economic model, which is based on a take-make-consume-throw away pattern. This model relies on large quantities of cheap, easily accessible materials and energy.

Also part of this model is planned obsolescence, when a product has been designed to have a limited lifespan to encourage consumers to buy it again. The European Parliament has called for measures to tackle this practice.

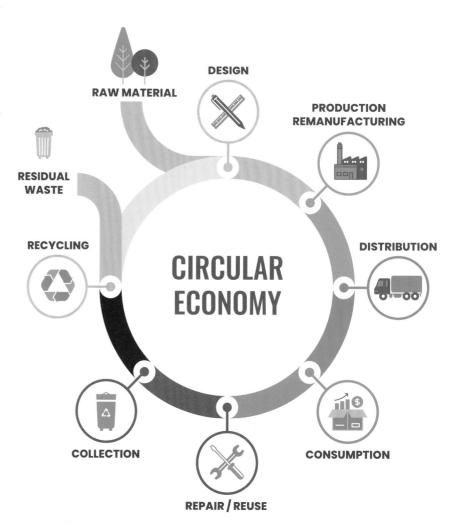

Benefits: why do we need to switch to a circular economy?

To protect the environment

Reusing and recycling products would slow down the use of natural resources, reduce landscape and habitat disruption and help to limit biodiversity loss.

Another benefit from the circular economy is a reduction in total annual greenhouse gas emissions. According to the European Environment Agency, industrial processes and product use are responsible for 9.10% of greenhouse gas emissions in the EU, while the management of waste accounts for 3.32%.

Creating more efficient and sustainable products from the start would help to reduce energy and resource consumption, as it is estimated that more than 80% of a product's environmental impact is determined during the design phase.

A shift to more reliable products that can be reused, upgraded and repaired would reduce the amount of waste. Packaging is a growing issue and, on average, the average European generates nearly 180 kilos of packaging waste per year. The aim is to tackle excessive packaging and improve its design to promote reuse and recycling.

Reduce raw material dependence

The world's population is growing and with it the demand for raw materials. However, the supply of crucial raw materials is limited.

Finite supplies also means some EU countries are dependent on other countries for their raw materials. According to Eurostat, the EU imports about half of the raw materials it consumes.

The total value of trade (import plus exports) of raw materials between the EU and the rest of the world has almost tripled since 2002, with exports growing faster than imports. Regardless, the EU still imports more than it exports. In 2021, this resulted in a trade deficit of €35.5 billion.

Recycling raw materials mitigates the risks associated with supply, such as price volatility, availability and import dependency.

This especially applies to critical raw materials, needed for the production of technologies that are crucial for achieving climate goals, such as batteries and electric engines.

Create jobs and save consumers money

Moving towards a more circular economy could increase competitiveness, stimulate innovation, boost economic growth and create jobs (700,000 jobs in the EU alone by 2030).

Redesigning materials and products for circular use would also boost innovation across different sectors of the economy.

Consumers will be provided with more durable and innovative products that will increase the quality of life and save them money in the long term.

What is the EU doing to become a circular economy?

In March 2020, the European Commission presented the circular economy action plan, which aims to promote more sustainable product design, reduce waste and empower consumers, for example by creating a right to repair). There is a focus on resource intensive sectors, such as electronics and ICT, plastics, textiles and construction.

In February 2021, the Parliament adopted a resolution on the new circular economy action plan demanding additional measures to achieve a carbon-neutral, environmentally sustainable, toxic-free and fully circular economy by 2050, including tighter recycling rules and binding targets for materials use and consumption by 2030.

In March 2022, the Commission released the first package of measures to speed up the transition towards a circular economy, as part of the circular economy action plan. The proposals include boosting sustainable products, empowering consumers for the green transition, reviewing construction product regulation, and creating a strategy on sustainable textiles.

In November 2022, the Commission proposed new EU-wide rules on packaging. It aims to reduce packaging waste and improve packaging design, with for example clear labelling to promote reuse and recycling; and calls for a transition to bio-based, biodegradable and compostable plastics.

24 May 2023

The future of recycling: innovation for the circular economy

The future place of recycling will be determined by improving waste management, developing new technologies, and incentivising reuse.

Recycling is one of those small steps for men, but a great step for mankind, on which the sustainability of the planet depends. Fortunately, recycling is increasingly embedded in society. Hundreds of projects and initiatives already exist which call on us to make the change toward a circular economy, where waste is transformed into resources with the aim of exploiting what we extract from the environment to the maximum.

What will I learn from this article?

* The importance of recycling

* Innovations in the area of recycling

* Classification of waste

* Treatment of waste

* Incentives to recycle.

The importance of recycling

We live surrounded by plastic: packaging, clothes, containers, tyres, suitcases, furniture, etc. Everything seems to be made from plastic or owes something to this material. But it has only been around for a relatively short while. It was back in 1950 when plastic began to be manufactured from petroleum and, since then, it is estimated that we have produced some 9.1 billion tonnes of the stuff.

Meanwhile, we have, at the most, recycled a tenth of all this plastic. The majority of it has been disposed of in waste tips and the natural environment.

Indeed more than 140 million tonnes of plastics now pollute the planet's rivers, oceans and lakes.

> 'The majority of plastic has been disposed of in waste tips and the natural environment'

This data comes from the Organisation for Economic Co-operation and Development (OECD), which in February called on the 193 UN countries to agree the first global treaty on plastics pollution.

> 'Of all the plastic that reaches the market every year, as little as 6% has been recycled'

In 2021 alone, production worldwide reached 461 million tonnes, an annual figure that has steadily increased over the past 20 years, says the OECD. And, of all the plastic that reaches the market every year, as little as 6% has been recycled.

The importance of innovation to recycling

The large quantity of waste we produce goes through a long process from its manufacture to, in the best of cases, transformation into recycled materials for new products. It is essential we find ways of managing waste intelligently, incentivising measures to promote responsible consumption and developing new techniques and processes that help citizens recycle more and better.

Step 1. Classification of waste

Materials can be recycled if they are clean and properly classified. This is much more complicated than it seems, since many people make mistakes when separating waste, or don't separate objects made from different materials before putting them into, for example, the yellow bins.

Thankfully, technology has already made progress in this field. Some waste selection plants employ infrared systems to classify different packaging and separate them according to the type of plastic they contain.

Thanks also to digitisation, the use of data and traceability of waste are becoming important. Now the waste can be tracked from its place of origin to its final destination, ensuring that it's treated correctly.

The use of data and traceability have the same goal: assure that the new material generated has sufficient quality to be reused by the consumer.

Step 2. Treatment of waste

There are already processes underway that employ high-powered microwaves to decompose polystyrene – a material we can find in many common objects, from packaging to electronic devices – into monomers. New polystyrene plastics can be produced from these molecules without the material losing quality or any of its properties.

The microwave process uses electricity instead of heat, making it a more energy-efficient recycling process which also considerably reduces the greenhouse gas emissions needed to produce polystyrene. The technology is so promising that one of the world's best-known tyre manufacturers has created a microwave-based tyre recycling system.

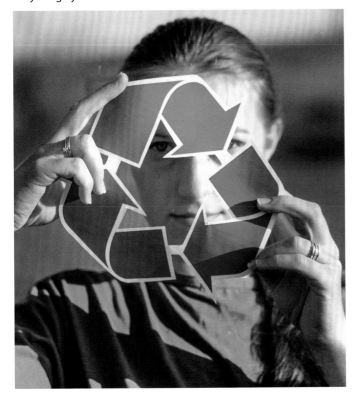

Step 3. Incentivizing recycling of waste

None of the above measures would be much good if we did not encourage people to recycle more and better. As said at the beginning of this article, recycling is a small gesture that starts in the home, but is also key to the sustainability of the planet.

New systems for returning and rewarding recycling seek to do precisely that: incentivize people to recycle more. A good example is a Spanish project that rewards citizens for correctly separating their waste.

Called RECICLOS, it consists of technology incorporated into containers and bins which allows people to connect with such infrastructure via their mobiles when recycling. It adds up how many times they have done so.

In this way, they can obtain rewards and exchange them for sustainable incentives like donations to food banks or prizes such as push scooters.

What is certain is that there are more and more facilities contributing to the growth of the circular economy and reuse of waste. It is in our hands to use them to care for the planet's natural resources and live in a more sustainable way.

Article by ACCIONA from Sustainability for all

Biodegradable shoes anyone? Four everyday items reimagined for nature

Whether it's seaweed packaging or bricks made from cow dung, a new crop of award-winning product designers are turning away from fossil-based plastics and chemicals.

By Jem Collins

Glance around your home and it won't be long until you spot something set to outstay its welcome. From ziplock bags to period pads, for almost everything we consume, a sustainable alternative is needed.

That's the goal of the Make It Circular Challenge, run by international organisation What Design Can Do. It called on entrepreneurs and creatives to submit project ideas that are built to last, work with nature, and use pre-existing resources. More than 650 entrepreneurs answered the call with innovative strategies for building a more circular society.

Here are four of the 13 inspiring winners.

1. Biodegradable shoes

David Roubach has been wearing the same pair of sandals for two and a half years now. It doesn't sound extraordinary until you account for the fact that they're completely biodegradable.

'I think it's funny that we as humans have invented so much, but for the end use of products, we only have one solution called recycling,' Roubach, the founder of Israel-based Balena, explains. He calls his innovation 'biocycling'.

Against the backdrop of the fashion industry, where 60 per cent of all material made into clothes is plastic, Balena has developed a bioplastic that ticks three boxes. Not only is it comfortable, sturdy and easy to mould, it's also made from

natural and renewable materials, and is entirely compostable in industrial facilities.

The company's first run of shoes, produced as a proof of concept, sold out, with customers able to return them for recycling when they're finished with them. But that's just the beginning. Roubach hopes to work on new materials, collaborations, and even certifications for products using Balena's materials.

But, he stresses, it all starts with consumers: 'The question we need to ask every time we buy something is: "what's going to happen to it at the end?"'

2. Seaweed packaging

Juni Sun Neyenhuys believes that packaging should only last as long as we need it. Whether it's a bag of salad leaves or a snack wrapper, once you've eaten the contents it should be thrown on the compost heap to biodegrade naturally and harmlessly.

That's the modus operandi of Germany based mujō, a company she co-founded with Annekathrin Grüneberg, which replaces conventional packaging with seaweed. Unlike land-based materials, seaweed doesn't need agricultural land, watering, or pesticides.

'All that bad stuff is not needed when you grow seaweed,' Neyenhuys explains. 'It has a positive impact on ecosystems because it [captures] CO_2, and it grows really, really fast. It can grow up to 60 centimetres in a day.

'[But] what is really important is that we don't make the same mistakes in aquaculture as we did in agriculture.' For example, harming biodiversity through mono-cropping, or relying heavily on synthetic fertilisers.

With funding now in place, the team is working on product development, and hopes to get its packaging on shelves within the next 12-18 months.

3. Plastic-free pads

'Our business model is the first of its kind,' Tarun Bothra, one of the co-founders of Saathi, based in India, tells Positive News. The company makes sanitary pads, but crucially, without the 3.4g of plastic that's usually in each one.

'It's a completely circular business model, where we are not taking anything at all which is manmade or from a resource that's depleting.'

Instead, the company uses banana and bamboo fibres, sourced from agricultural waste. Antibacterial properties of these materials make them a perfect fit for period products, and the pads take only six months to decompose after use.

Saathi has also donated over a million of its pads to Indian women – only 36 per cent of whom have access to menstrual products.

'We're trying to propose a more holistic approach to menstrual hygiene, as well as health and sustainability,' says co-founder Kristin Kagetsu. While there's a long way to go, she believes a shift in mindset is taking place.

'We've seen fathers come and talk about the product with their daughters, we've seen guys buy it for their friends. That's definitely a new experience for us.'

4. Bricks made from cow dung

CoolBricks are different to other building materials – but you wouldn't know it. The first reaction of masons when given a brick to hold is that they can't believe it doesn't contain cement, says founder Emile Smeenk. Unlike everyday bricks, these ones are not fired or baked and are 100 per cent recyclable.

'CoolBricks are composed of soil and the active ingredients of cow dung,' explains Yask Kulshreshtha, head of research and development at the startup, which is based in the Netherlands and Uganda.

'Traditionally, many countries have used cow dung for making houses,' she continues, explaining that her team has taken this core idea, and modernised it.

While the mud is often cut from wetlands, leading to habitat loss, CoolBricks' method can incorporate any type of soil as the base ingredient. It also pays farmers a fair price for their cow dung.

The result is a brick that is 20 per cent stronger than traditional mud bricks, cuts carbon emissions by 90 per cent and costs an impressive 50 per cent less.

'Circular innovation is about shortening supply chains and making it simple so that you can reuse local materials,' says Smeenk.

19 July 2023

Design

Take a look around your home or classroom. Can you identify any items or products that could be improved in the way they are made? Choose one thing and create a sketch showing how you would alter its design and the materials you would use to make it more sustainable.

What is circular fashion?

Circular fashion ditches the linear 'take-make-waste' model and instead asks the industry to close the loop on production, including responsible manufacturing, use, and end-of-life for every garment.

By Madeleine Hill

Circularity in fashion to close the loop

Remember the times when words like 'sustainable', 'ethical', and 'eco-friendly' were rarely used in the public domain? In recent years, as highlighted by ongoing issues like the climate crisis and wage theft, the significance of these terms has skyrocketed. One in particular, a relative newcomer to the block, is 'circular fashion'. But what is circular fashion, and how can it help?

Considering the strong presence of these new concepts and what they represent, and as shoppers and consumers ourselves, it's hard not to wonder–where do we fit into all of this? For conscious consumers who support sustainability in both fashion and other areas of their lives, we are increasingly seeing that this new categorisation is worth exploring.

In this article we will dive deeper into what circular fashion is today, how it came to be, and what we can do to align with the concept in our consumer choices–not only in relation to fashion, but in our lives more broadly.

Circular fashion is...

Circular fashion is a system where our clothing and personal belongings are produced through a more considered model: where the production of an item and the end of its life are equally as important. This system considers materials and production thoughtfully, emphasising the value of utilising a product right to the end, then going one step further and repurposing it into something else. The focus is on the longevity and life cycle of our possessions, including designing out waste and pollution. Essentially, the 'circular' comes as a response to previous economic and societal models that have been 'linear' to date, and harmful on the planet along the way.

Circular fashion is a system where our clothing and personal

belongings are produced through a more considered model: where the production of an item and the end of its life are equally as important.

Further to this, circular fashion comes from the collision and intersection of the 'circular economy' (a model that exchanges the typical cycle of take-make-waste in favour of as much reusing and recycling as possible) with sustainable and ethical fashion. The development and evolution of these two areas ran parallel for some time within different sectors. This new category is a much needed addition to the fashion industry's sustainability journey and progression, particularly as it brings stronger ambition and advocacy, as well as a commitment to investing in clothing that will last far longer than fast fashion counterparts.

Key points of circular fashion:

- Using less materials when producing individual items for increased recyclability

- Working to remove nonrecyclable and polluting materials from the supply chain

- Recapturing everything from garment offcuts to packaging for reuse

- Ensuring use and reuse for as long as possible including collection schemes and bringing the recycled materials back to a 'good as new' state

- Returning any unavoidable waste to nature safely

How circular fashion came to be

It was around 2014 when fashion first officially collided with the circular economy, resulting in the newfangled term of 'circular fashion'. The term was first coined at a seminar in Sweden, where a more circular approach to the fashion industry was the core focus. This pivotal distinction came at a time when the floodlights were cast on an industry whose impact was coming under serious scrutiny. Only a year before in 2013 the Rana Plaza clothing manufacturing complex in Bangladesh collapsed, killing over 1,000 workers, with fast fashion quickly becoming an undesirable model for consumers to support. This tragedy, along with the last few years especially bringing to the forefront some questionable methods in the industry–including devastating environmental, human, and animal impacts– have also highlighted the important role consumers play in advocating for better standards and in fast-tracking change.

Since 2014, the rigorous shift to support more sustainable, ethical, and circular systems has increased tenfold, and has been present in numerous industries in relation to how we as a society can act, choose, and do more sustainably. At the core of this revelation is knowing where things come from, what they are made of, who made them, and being accountable for the overall lifecycle of our belongings. In reality, it is often hard to imagine a time when sustainability and all these considerations weren't an expected consideration (even if only on a small scale).

As circular fashion has developed in the last few years, an interesting part of this model is what it represents. We know the circular economy has been around in academics and business for many years prior to clothing and production, with an increased appetite for this model globally. But, looking forward, the question for fashion remains: does this convergence of more sustainable and circular models mean a whole new economic fashion system? As consumers align with this model by buying less, owning items for longer, and being more specific in their choices (essentially slow fashion), can circularity co-exist with sustainably produced fashion?

One answer could be that we may see an overall shift in our economic structure to a baseline standard of better made products that consumers will invest in for longer periods of time. Simultaneously, we will hopefully see continued support for slow fashion (consuming less) which will result in a cleaner, healthier, less impactful industry overall. In this equation, there are important roles for both consumers and the larger corporations to play for us to see an overhaul of the fashion industry's significant environmental footprint.

The role of consumers

Consumers play a key role in the development and implementation of circular fashion. From increased demand and support for more sustainable models, we are seeing circularity seeping into supply chains, manufacturing, and at the final stage; after a consumer has finished with an item. With more stores offering recycling programs and councils looking at textiles recycling, there are more and more options and alternatives for the materials that make up our clothes, rather than falling straight into landfill, which is great news for the planet (and for your pocket).

Some key actions consumers can take to align with circular fashion:

- Know more about the brands you buy by using resources such as Good On You
- Support more sustainable and ethical fashion
- Live by the five r's of fashion
- Buy less and buy better
- Revise your wardrobe before buying new
- Shop second hand where possible
- Consider renting for your next event
- Host and attend clothes swaps
- Look after your clothes
- Question 'What are the alternatives?' before throwing away used clothing
- Utilise in store recycling programs
- Choose more sustainable materials when purchasing new clothing
- Talk to people about the benefits of circular fashion
- Learn the impact our clothing has on the planet and the importance of where your clothes are made, who made them, and what they are made of
- Make a commitment to not buy brands who don't strive to sit within the circular model

Circular fashion has brought a wave of greater consumer knowledge, powerful advocacy, and overall acknowledgement that previous, linear approaches to the fashion industry can't continue. This demand for transparency, longevity, and a new framework is set to continue well into the future, and represents a future for fashion that would be less impactful and more in harmony with all the resources, processes, and people involved.

Although we are a long way from a completely circular model in fashion, as more and more brands and consumers become aware of and invest in it, we are already preventing waste and degradation that would otherwise exist, which is a step in the right direction.

19 September 2022

Fashion's sustainability reckoning

Clothing businesses are bracing for new rules that could upend their business model.

By Leonie Cater

COPENHAGEN – On the surface, it was everything you would expect from a high-end fashion industry knees-up.

Green juice for breakfast, impeccably dressed conference-goers, Moët & Chandon flowing before dinner.

But the topic at the heart of this year's Global Fashion Summit in Copenhagen was less glamorous: How can a sector that has thrived on novelty and extravagance survive global efforts to slash carbon emissions and eliminate waste?

From bringing more transparency to supply chains to developing more sustainable manufacturing methods, panels revolved around how the fashion industry can fall in line with upcoming EU and U.S. regulations – sharp-ish.

That's with good reason. The bloc's textiles industry faces an onslaught of regulation that could force a reckoning over its environmental and human rights abuses.

The EU's textile and clothing sector had a turnover of €147 billion – and trillions globally – in 2021, according to industry body Euratex. All of that economic activity has caused significant environmental damage, including chemical pollution caused by viscose factories, mountains of textile waste and a hefty carbon footprint.

Some 80 percent of those impacts occur beyond the bloc's borders, where most textile production takes place, according to the European Environment Agency. Cotton farming, fiber production and clothes manufacturing, for example, mostly happen in Asia.

Workers' rights are another sore spot. According to estimates by the Clean Clothes Campaign, the minimum wage for garment workers in Bangladesh is $94 per month, while the living wage is estimated at $569 per month. Ten years after a factory in Bangladesh collapsed, killing more than 1,100 garment workers, in April a fire blazed through a garment factory in Karachi, Pakistan, killing four firefighters and injuring nearly a dozen others.

And after all that, in Europe, a garment only ends up being worn an average of seven or eight times before being tossed.

Regulatory onslaught

In a bid to clean up the industry's mess, the EU is moving forward with new rules on supply chains, greenwashing and sustainable design.

New EU ecodesign rules, for example, would force companies to abide by (as yet undefined) rules on the sustainable manufacture of their clothes. A ban on companies destroying

unsold goods is on the horizon and, just this week, Brussels announced new laws forcing the industry to pay for the clean-up of the waste it produces.

Most of these proposals won't enter into force for several years, but the change for the industry could be costly.

At the Global Fashion Summit in the Danish capital, speakers and conference-goers put on a brave face.

Nicolaj Reffstrup, co-founder of Danish fashion brand Ganni, said his brand welcomes "any kind of legislation that will kind of level the playing field and push the agenda forward."

Amanda Tucker, vice president of responsible sourcing and sustainability at U.S. retail giant Target, stressed the importance of "involving brands and suppliers in the formulation of these policies" or it "won't be implementable."

But industry players are also worried about the increasingly uncertain future.

"We are an unregulated industry so far, so everything is new and people are scared," said Clémence Hermann, senior manager of public affairs and sustainability at online fashion portal Zalando. "People are scared of uncertainty, of change. People don't know where to start."

Turning up the pressure

Faced with the threat of more regulation, companies are rushing to work out how to police their own supply chains and green their manufacturing processes.

TrusTrace, a software company that claims it can help companies monitor for red flags in their value chains, said it has seen business skyrocket in the past couple of years.

Alternative leather brands like Desserto — which creates leather-like materials out of cacti and plastic — are raking in the industry collaborations.

But for green campaigners, things aren't moving fast enough.

Valérie Boiten, a senior policy officer at the Ellen MacArthur Foundation, said schemes forcing fashion companies to take responsibility for the waste produced by their products is a "significant first step" but it will "not disrupt the short lifetimes of our textile products."

"It's amazing that fast fashion brands or fashion, for the first time, are responsible to pay for that. But it's the minimum and you're talking about gigantic businesses," said Livia Firth, co-founder of sustainability consultancy Eco-Age. "That, for them, is like peanuts."

What's needed, Boiten added, is a "radical transformation in terms of circular product design and business models to ensure textiles aren't discarded in the first place."

Common demands among green textile campaigners include concrete EU targets for the reuse, recycling of textile waste and strong measures in Brussels' upcoming ecodesign laws for the textile industry.

They're backed by the European Parliament, which earlier this year overwhelmingly voted to push the Commission to include targets for textile waste prevention, collection, reuse and recycling in the upcoming legislation.

Green MEP Anna Cavazzini said the EU urgently needs to "reduce our consumption of materials, set objectives for waste prevention, and facilitate the reduction of carbon emissions in production and products."

Fashion leaders say they're listening — but warn that the revolution won't happen overnight.

"I hear the urgency and [they're] totally right but how do we actually make it happen in a way that's actually doable?" said Zalando's Hermann. "We need to start from very low. And it's going to take time. But I think people are committed."

7 July 2023

This Article was first published by POLITICO and written by POLITICO reporter Leonie Cater. No changes have been made.

Stop 'wishcycling' and get wise: how to recycle (almost) everything

From contact lenses to blister packs and used dental floss, there are items that perplex even the most dedicated recycler. Here is the expert guide to getting organised – and getting rid of your rubbish.

By Emma Beddington

I spend a lot of time – too much time – thinking about recycling and the main thing I think, over and over, is: it shouldn't be this hard. Eighty per cent of UK households are 'still unclear' about how to recycle effectively, according to research last year – and who can blame us?

Labelling often requires a doctorate in semiotics to decode, kerbside collections are a postcode lottery and council recycling centres are often difficult to access without a car. At home, packaging piles up – no one knows what to do with toothbrushes or the cat's treat packages, and we're squabbling over pizza boxes. All of it amounts to us collectively wondering whether recycling is ultimately pointless because it's all going to end up in landfill in the developing world.

'We've made something that could be fairly simple really complicated,' says Libby Peake, of the environment thinktank Green Alliance. 'And that's quite frustrating for the public who want to do the right thing when it comes to recycling.'

Things are changing, though, if much too slowly. 'We have come a long way,' Steve Eminton of industry news website Let's Recycle reminds me. 'Ten years ago, places wouldn't recycle milk bottles or yoghurt pots.' The findings of a government consultation on 'consistent collections' are due imminently and the hope is that eventually this will mean not just plastic, glass, paper and card but also cartons (Tetra Paks and similar) will be collected from all homes.

> The message of 'reduce, reuse, recycle' has been lost – we focused on the thing that is the least important

There's also the proposed deposit return scheme, under which consumers would pay a returnable deposit for plastic bottles and cans (and glass in Wales), but, in most of the UK, not until 2024 at the earliest. (Scotland is due to launch its scheme, which will also include glass, in August 2023). Eventually, technological advances will make a difference too: AI-enabled sorting, apps that allow you to scan packaging before you bin it, and a watermarking system for materials are all in the works, according to Archipelago, a fund investing in solutions for hard-to-recycle stuff.

But in the meantime, what can we do? For a start, maybe fixate less on recycling. That's an odd thing to say in an article about recycling, but it's supposed to be a last resort: limiting waste has more impact and reuse is a better strategy where possible. As Peake says: 'The messages of the three Rs (reduce, reuse, recycle), I think, have been lost in a lot of ways, and we focused on the thing that is the least important but probably the easiest to grasp.'

Next, don't 'wishcycle': putting stuff in the recycling because you wish, or hope, it could be recycled causes more problems than if you chucked it in the bin. It's also important to maintain pressure on manufacturers, retailers and government: things won't improve unless we show we want them to.

Subject to these caveats, here is a guide to what to do with some of the household items we struggle with, or don't know how to recycle. If you're still unsure, the Recycle Now website is a godsend. Managed by waste charity Wrap, it provides a guide to what can and can't be recycled in your postcode.

Contact lenses and packaging

The depressing mountain of foil-topped plastic blister packs from my sons' daily contacts launched me on this quest. There are solutions: Specsavers collects contact lenses and lens packaging in every store and recycles them in the UK with a company that turns them into construction materials. Boots Opticians will also take back lenses and packaging but theirs is recycled through TerraCycle. It's fair to say the recycling world is agnostic at best when it comes to TerraCycle: BBC *Panorama* has reported on issues with the company's supposed UK recycling ending up in Bulgaria or

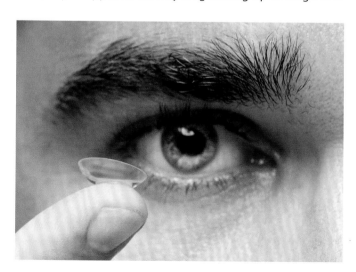

left to pile up with subcontractors, or possibly ending up in landfill in the US. I have suggested alternatives where they exist.

Glasses

Lions International has recently expanded its scheme collecting spectacles at its Birmingham HQ, then partnering with charities to get them to eye centres and clinics in the developing world (currently the Gambia, Nigeria, Chad, Bangladesh and Mali). You can find out how to donate by emailing enquiries@lionsclubs.co.uk or calling free on 0345 8339502. Alternatively, Specsavers, Boots Opticians and Peep Eyewear take glasses for recycling.

Cosmetics

Mascara, lipstick, makeup palettes, travel miniatures ... Boots takes all of these in store, from any brand or source. They go to a UK-based recycler to be transformed into construction board. Boots guarantees nothing goes to landfill and nothing is incinerated.

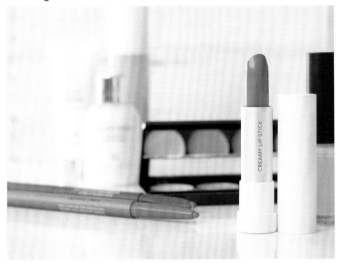

Pumps

The tops from soap, shower gel or other dispensers can't go in your normal plastic recycling. Ideally, refill and reuse pump bottles as far as possible, but the Boots cosmetic recycling scheme also takes them.

Dental stuff

When it comes to toothbrushes, electric toothbrush heads, floss containers and those interdental brushes dentists love, there isn't much good news currently. TerraCycle will accept them if you can find a dropoff point, with the reservations explained above. Toothpaste tubes can go in Boots cosmetic recycling boxes. For the rest, 'reuse them as much as you can for cleaning or something', advises Eminton. Conventional dental floss can't be recycled, but there are now silk or plant-based flosses (I use a corn starch one) and glass dispensers that can be refilled indefinitely.

Insulin pens

Boots and Superdrug take these through the PenCycle scheme.

Pill blister packs

This is tricky. 'Because it's pharmaceutical, it's very stringent on what plastics you can use, so a lot of them are PVC and, from a recyclability perspective, that's very difficult,' says Adam Herriott of Wrap. Superdrug runs a blister pack recycling scheme in stores where it has an in-store pharmacy – the store locator allows you to filter to find one. Returned packs are turned into boards for the construction industry.

Some people report difficulty in having their blister packs accepted in stores: Superdrug says the scheme is still live, but it doesn't have the capacity to cope with large-scale community collections, only individual drop-offs, which may be the issue. There may be an alternative local scheme in your area: a group of GP practices is running a pilot in mine.

Black plastic containers

Finally, some good news. In the past, black plastic takeaway and ready meal containers could not be recycled. 'It's not an issue at all nowadays,' says Herriott. Even something that appears black is unlikely to use the problematic carbon that was the barrier to recycling in the past, and recycling equipment is perfectly able to deal with all widely used pigments these days. Despite this, some local authorities still do not accept black plastic, so check before your chuck, but most of us should be able to recycle away.

Bags and wrappers

One of the big innovations of recent years has been the front-of-store collection boxes for flexible plastic in supermarkets. They take salad bags, carrier bags, crisp and biscuit packets, ready meal film lids and more: if it springs back when you crumple it, it can go in there.

At Co-op, recycling takes place in the UK, where material is sorted by polymer type and made into bin bags, rigid storage items and our old favourite, construction boards. One key point to make about this stuff: please do rinse, as food soiling is likely to make it impossible to recycle.

Pizza boxes

Eminton puts this one to bed for me: 'They can be recycled unless they're really dirty – it's just common sense.' Grease stains are fine.

Food and pet food pouches

All of these can go into the flexible plastic collection points in supermarkets, though it's worth highlighting that they are among the most difficult plastics to recycle, because of their multiple layers. So they become – you guessed it – construction boards. Pets at Home will also recycle pet food pouches of any brand in its 320 stores – it asks that you rinse them out first.

Foil

It's worth checking on the RecycleNow website or directly if your council collects foil kerbside – I discovered mine does. Otherwise, your local recycling centre will probably take it. Scrunch before you recycle, ideally creating something tennis-ball-sized or larger.

Coffee pods

There are so many more ecological ways to drink coffee, but if for some reason you're tied to pods like Nespresso, recycling options have improved recently. Podback offers a kerbside service in nine local authority areas (check on the website). Other customers can pick up free recycling bags from Morrisons, Ocado or one of the participating pod brands and send it back using Yodel's Collect+ service.

Jar lids

Keep metal jar lids in a tin and either put into your tin can recycling kerbside, or in the metal recycling at your local waste facility.

Compostable' or 'biodegradable' plastic wrappers

The tide is turning on these plant-based plastics, which were initially heralded as a revolution. You may have spotted the results of University College London's *Big Compost Experiment* published late last year highlighting how poorly compostable they actually are. Wrap's advice? Don't try to home compost these, and check what your local authority says. If there isn't a food waste collection, says Herriott, it's unlikely your council will collect compostable plastics and they should just go in the bin.

Tetra Paks and other cartons

Although cartons are recyclable, the multiple layers mean it's a fairly complex process and they can't be recycled into more cartons (known as 'closed loop' recycling). So if a plastic bottle is available for the same product, such as milk, choose that instead. Where you can take them is up to your local council. Check the Recycle Now website to see whether they're picked up kerbside, or if you have to drop them off somewhere.

Cookware

Even with a non-stick coating, metal or mostly metal pans should be fine to go in your nearest council metal recycling point. 'Teflon coating shouldn't make any difference,' says Eminton.

Padded envelopes

A paper padded envelope can be recycled with your card and paper, according to Eminton. Plastic will need to go in the bin.

Polystyrene packaging material

If you're lucky enough to live within the north London waste authority, you can take your polystyrene packaging to your local tip. For the rest of us, there isn't a good solution, unless whoever delivered it will take it away again. 'Better to put it

in your black sack than put it in your recycling bin, because that causes a problem,' says Eminton. There's a lot less of it around nowadays, thankfully, and it's also widely recycled by commercial users.

Textiles

Although there are plenty of well-known ways to give your old clothes away – charity shops, collections for asylum seekers or the H&M scheme that takes back unwanted clothes – there's no great news here: textiles are still poorly recyclable. 'A jacket has a fleece lining, an inner material, an outer material, buttons, zips and threads – we can't separate all of that,' says Lucy Mortimer of Archipelago.

'Even if you can deconstruct the individual components, you then have a problem with the individual materials,' adds her colleague Justin Guest – and technology to separate fibres is still in its infancy.

So what can you do? You already know: buy less and buy secondhand. But also, be sceptical around claims about recycled material in new clothes: plastic bottles are better turned into more plastic bottles than clothes that then can't be recycled.

Teabags

Loose-leaf tea is the best option for the planet (all that packaging significantly ups the carbon footprint), but if it has to be bags, choose one with a plant-based seal. Unlike oil-based plastic seals, these are made from renewable resources and can be composted industrially.

Many bags are now sealed with plant-based plastic, including PG Tips, Yorkshire Tea, Clipper, Co-op, Asda and Sainsbury's own brands – Tesco is moving across by summer of this year.

If your local authority collects food or garden waste, put them in there; if not, you can put them in home compost, though you may end up having to retrieve small amounts of plant-based plastic residue. The super-virtuous empty the leaves into compost and bin the bag (also the best approach if you're not sure what your teabag is made of).

Cables and chargers

The total number of cables hoarded in UK homes that could be recycled (140 million) could circle the Earth more than five

times, according to Material Focus. Its website, Recycle Your Electricals, tells you what to do with cables and everything else. 'Anything with a plug, cable or battery can be recycled,' says Scott Butler, the organisation's executive director. His advice? Bag them up and when you've got enough use its online postcode locator, which already has 5,000 recycling points and is adding more constantly.

CDs and DVDs

You can't recycle DVDs at home, but places such as Zapper will take them, and might even give you a bit of money for the good ones.

Vapes

Every week, 1.3 million single-use vapes are thrown away, according to recent research and they are the UK's fastest-growing waste stream. They don't need to be: follow the Recycle Your Electricals advice and find a recycling point. Take the battery out if there is one and recycle it separately; don't put the whole vape in with the battery recycling. Also consider switching to refillable vapes – many e-liquid bottles are made from recyclable plastic (but check if unsure).

And what can't you recycle?

Large plastic toys

There's no recycling route for garden slides and other large plastic kids' items, but given they are also almost indestructible, share the love by giving them away.

VHS cassettes

Short of creating some kind of retro hipster sculpture with them, your videotapes need to be binned.

Pyrex dishes and drinking glasses

Don't put these in your glass collection: the different melting points cause havoc with glass recycling.

Disposable nappies

There's no large-scale or widely available solution for the stinky stuff but nappy recycling pioneers Nappicycle in Wales have resurfaced the A487 between Cardigan and Aberystwyth with them. Partnering with Nappicylcle, babycare company Pura has also just completed a six-month trial in Bristol, so watch – don't sniff – this space.

7 February 2023

The Future?

Sustainability is 'key to unlocking £70 billion economic opportunity' for UK

By George Heynes

The UK has a 'once in a generation economic opportunity' to unlock £70 billion by becoming a 'sustainability superpower', according to a new report.

A report released by the UK Business Council for Sustainable Development (BCSD) has found that the economic opportunities for the UK to adopt a 'beyond net zero' strategy could see the UK established at the forefront of the renewable market.

The report highlights the opportunities the UK has in becoming a major exporter of renewable energy, referencing that it could be similar to when the UK became a major exporter of North Sea oil and gas during the 1980s and 1990s.

Offshore wind could be pivotal in achieving this aim. With the nation's advantage in being situated in close proximity to the North Sea, the Atlantic Ocean and the Celtic Sea, there are vast areas that could be used to generate large quantities of renewable energy via offshore wind.

In light of this, the UK Government has already established a 50GW by 2030 target for offshore wind. Through this, it is hoped this will support its decarbonisation goals and attract investment within the sector. It has also introduced key legislations to help achieve this target such as the widely successful Contracts for Difference (CfD) scheme.

However, despite the positives of the UK Government, there are major barriers preventing the UK from becoming a clean energy superpower which must be overcome to help the nation secure the economic opportunity.

These obstacles, which could become enablers if overcome in the correct way, include:

- Creating a National Grid that is fit for purpose.
- Delivering massive energy storage, using a range of technologies, to capture the tens of Terawatts of power lost each year when Britain needs to switch off renewable energy sources because they are generating more energy than the grid can handle.
- A national programme of retrofit to fix Britain's stock of old draughty homes, offices, and other commercial properties.
- Early intervention to catalyse the market for hydrogen.

According to the report, in the 'beyond net zero' scenario, the UK could achieve 'economic benefits of £70.3 billion a year by 2050 – not including the social value of reduced greenhouse gas emissions and averted climate change. This includes an additional £36.4 billion of Gross Valued Added (GVA) delivered by clean energy generation and a £17 billion boost to the UK's trade balance. There would be a further £2.2 billion of agglomeration benefits, while disposable incomes would see a £14.7 billion bump as a result of lower energy prices.'

Alongside offshore wind, the report also outlines a number of other technologies that need to be supported to help achieve the beyond net zero scenario such as solar, energy storage and hydrogen. The report includes a number of potential policy options to help fully exploit the UK's clean growth potential. These include:

- An annual quota of government interest-free 'retrofit loans' provided to owner-occupier households, as well as residential and commercial landlords.
- Large-scale rooftop solar installations to achieve a five-fold increase in solar capacity by 2035.
- Government innovation funding to leverage private investment, for the development and trialling of alternative storage solutions to hydrogen and electric batteries.
- Use of 'priority grid connection auctions' for electricity generators and energy storage providers to raise money for investment in National Grid priority projects.
- Regulating to place an obligation on wholesale or retail gas suppliers to blend all gas supplies with a residual amount of hydrogen.

'Today, more than 90% of global GDP is covered by some form of net zero target. The findings from our report are clear. The UK can unlock more than £70 billion of economic benefits a year if we become a world leader in the race to net zero,' said Jason Longhurst, chair of the UK Business Council for Sustainable Development.

'We have the potential to generate huge amounts of clean energy which would turn the UK from a net importer of energy to a nation exporting vast amounts of clean power, worth £17 billion a year, to mainland Europe.

'We believe this paper delivers an evidence base to enable our government to drive new incentives to transition, leverage in further private sector investment and position the UK as one of the world's most investable markets for companies tackling the challenges created by climate change.

'Having fired the starting pistol on the race to net zero it's now time for the public and private sector to work together to put Britain at the front of the field again.'

15 May 2023

DWP Estates outlines sustainability plans

This article sets out DWP's carbon and water sustainability plans to 2025.

DWP (Department for Work and Pensions) Estates has published its carbon and water sustainability plans for the next 3 years in line with the government's greening commitments (GGCs). This will be reviewed once the new GGCs are published in 2025.

As a response to the Committee on Climate Change (CCC), individual government departments must publish their plans on cutting emissions targets and reducing greenhouse gas emissions.

DWP Estates has already put in place a large range of activities, strategy measures and policy changes in order to meet the 2024 to 2025 Greening Government Commitment and longer-term targets. These touch on almost every area of DWP Estates, which has a mission to be smaller, better and greener

DWP's GGC climate targets for 2021 to 2025 are as follows:

- reduce overall greenhouse gas emissions by 45% from a 2017 to 2018 baseline and reduce direct greenhouse gas emissions from estate and operations by 17% from a 2017 to 2018 baseline

- develop an organisational Climate Change Adaptation Strategy across estates and operations

- conduct a Climate Change Risk Assessment to better understand and to target areas that need greater resilience

- develop a Climate Change Adaptation Action Plan, including existing or planned actions in response to the risks identified

- reduce the overall amount of waste generated by 15% from the 2017 to 2018 baseline

- reduce water consumption by at least 8% from the 2017 to 2018 baseline

DWPs wider Sustainability targets for 2021 to 2025 also include:

- investigating Green Leases and how we engage with landlords on delivering energy reduction projects

- establishing Operational Energy Use Intensity requirements for new build and major refurbishment projects

- carrying out Low or Zero Carbon energy studies for new build and major refurbishment projects

- revising processes so that sustainability is always factored into asset and lifecycle decision making

- ensuring that Building Management System (BMS) continue to be improved, ensuring our estate is managed in the most efficient way

- initiating improvements to energy and water management in our current facilities management contract, and enshrining this across any new contracts

- seeking continued improvement and enhancement to the Estates Design Standards

- continuing to work with internal Sustainability Champion volunteers to foster staff behavioural change

- collaborating with other areas of the department to ensure that the new targets are understood and being addressed in their area

- developing Nature Recovery Plans incorporating land, estates, operations and resources, and existing biodiversity plan

- seeking improvements to our waste management including reducing paper and plastics, and improving recycling:

  - continue to buy more sustainable and efficient products and services with the aim of achieving the best long-term, overall value for money for society

  - develop and deliver Nature Recovery Plans for their land, estates, developments, and operations

  - report on the adoption of the Greening Government: ICT and Digital Services Strategy

20 June 2023

Beyond growth: a new path for a sustainable future

By Charlie Malcolm-McKay

As humanity continues to grapple with the growing impact of climate change, many are questioning the traditional economic model of perpetual growth. Increasingly, experts and thought leaders are discussing an alternative approach – a radical and groundbreaking concept that challenges us to reevaluate our priorities. This idea envisions a world where the quest for profit takes a backseat to the development of resilient, equitable, and environmentally responsible societies.

The case for degrowth

At its core, the concept of degrowth is a vision of a future that is less centred on endless consumption. It is an economic theory that advocates for a radical reduction in contemporary consumerism, with the aim of achieving a sustainable economy. Not only that, but it is also a call for us to rethink our relationship with the natural world and to prioritise the well-being of all people, not just a privileged few. The concept emerged in the 1970s as a response to the lack of compatibility between capitalism and the survival of the planet. It has since grown into a global movement, with proponents calling for systemic changes that address the root causes of ecological destruction.

Advocates of degrowth argue that continuous economic growth will deplete our finite planetary resources, causing irreversible damage to the environment. For example, according to the United Nations, global resource extraction has tripled since 1970. The rate is expected to reach 190 billion tons of resource extraction by 2060. In contrast, degrowth advocates for a steady-state economy where resources are used at a sustainable rate, and the focus is on achieving well-being rather than maximising profit. To achieve this, proponents of degrowth argue for a radical restructuring of society, including a shift away from fossil fuels, a reduction in consumption, and a focus on community-oriented activities.

One of the key benefits of degrowth is its potential to reduce greenhouse gas emissions and mitigate climate change. According to a report by the Intergovernmental Panel on Climate Change (IPCC), reducing consumption and shifting to sustainable lifestyles are crucial for limiting global warming to 1.5 degrees Celsius above pre-industrial levels. By reducing the production and consumption of goods and services, degrowth could significantly reduce greenhouse gas emissions and contribute to the fight against climate change.

Additionally, degrowth could lead to increased social equality by reducing the gap between the rich and the poor. According to the World Inequality Database, the top 1 per cent of the world's population owns more wealth (38 per cent of total wealth) than the bottom 50 per cent (2 per cent total wealth). Proponents of degrowth argue that a reduction in consumption and a focus on community-oriented activities could help redistribute wealth and reduce income inequality. For example, a community centred on degrowth principles could decide to prioritise local businesses and services over large corporations. This would involve supporting small businesses in the community, encouraging local entrepreneurship and reducing the amount of money that flows out of the community to large corporations. This could help to redistribute wealth and reduce income inequality by creating more opportunities for people in the community to start their own businesses and share resources.

Obstacles to degrowth

However, the concept of degrowth is not universally accessible and has faced criticism from those who argue that economic growth is necessary for developing nations to improve healthcare, education, and other essential services. According to a report by the World Bank, economic growth is essential for reducing poverty and increasing access to basic services such as clean water, healthcare, and education. Developing nations, in particular, may not have the luxury of choosing a degrowth model and may need to focus on achieving economic growth to improve the standard of living for their citizens.

Despite these criticisms, proponents of degrowth argue that the current economic model is unsustainable and that the costs of inaction are too high. In a report published by the Sustainable Europe Research Institute, the authors argue that: 'societies that live within ecological limits and social boundaries are more resilient and better prepared for future shocks and crises.' For example, The Bishnoi community from the Thar Desert in India has been practising environmental conservation and sustainable living for over 500 years. Recently, the community has faced a severe water scarcity crisis due to droughts and overexploitation of groundwater. However, their traditional water harvesting methods and efficient use of water resources have enabled them to cope with the crisis better than other communities in the region. By shifting towards a degrowth model, societies can create a more sustainable future and avoid the worst impacts of climate change.

So what does a society centred on degrowth economic principles actually look like? In this society, infrastructure investments would focus on renewable energy sources, public transportation, bike lanes, pedestrian walkways, green spaces, sustainable building materials, and waste reduction systems. While prioritising sustainability, these investments could potentially create new job opportunities

in the renewable energy sector and contribute to the health of the environment.

One potential downside could be the cost of these infrastructure investments. Transitioning to renewable energy and sustainable building materials may be more expensive in the short term. There is also the issue of mining and extracting minerals required for renewable energy infrastructure, such as rare earth metals for electric cars. This often leads to land dispossession where Indigenous communities are forcibly removed from their ancestral territory to make way for extraction activities. This can lead to the loss of livelihoods and identity. For example, in Scandinavia, increasing mining pressure is being placed on the Indigenous Sami livelihood of reindeer herding. Increasing extraction activities disrupt the reindeer's traditional grazing grounds and threaten the Sami's cultural heritage.

Changing how we do things

Despite these obvious challenges, introducing degrowth as a viable economic principle could shift education towards developing skills and knowledge that contribute to a functional and sustainable society. This could mean investing in environmental education and sustainable agriculture, as well as in skills such as conflict resolution, communication, and community organising. Following this pattern, the job market would also shift towards more socially and ecologically valuable work. This could mean investing in care work, such as healthcare, as well as in ecological restoration and conservation efforts. While it's true that transitioning away from industries such as fossil fuel extraction and certain forms of manufacturing could lead to job loss and economic instability, it's crucial to consider the long-term consequences of continued environmental attrition.

In today's rapidly changing world, the idea of degrowth has emerged as an economic theory that provides a new path towards a sustainable future. While there are real benefits to this model, it is not without its challenges, particularly for developing nations. However, the urgency of the climate crisis demands bold action, and degrowth is one potential solution to create a more sustainable future. As Jason Hickel, an economist and leading proponent of degrowth, notes: 'degrowth is not a choice, but an inevitability if we want to avoid ecological collapse.'

12 April 2023

Key Fact

- According to the World Inequality Database, the top 1 per cent of the world's population owns more wealth (38 per cent total wealth) than the bottom 50 per cent (2 per cent total wealth).

Write

Write a short paragraph explaining the concept of 'degrowth'. Include an example that illustrates the idea.

E-waste to secure a sustainable future

Fredrik Forslund, VP Enterprise and Cloud, Blancco, argues that e-waste mustn't be overlooked in our efforts to secure a sustainable future.

In recent years, sustainability and sustainable practice have evolved from a 'nice-to-have' to a significant priority for enterprises, public sector organisations and governments. It has risen on the agenda as organisations look to tackle issues of wastefulness and declining biodiversity. This shift is not just being driven by enterprise customers and consumers; investors are now pushing companies to reduce their impact on the world around them.

Towards the end of 2020, the UK Government outlined its 10-point plan for a green industrial revolution. In June 2021, more than 450 investment firms also signed a letter outlining five courses of action governments around the globe should take to improve climate-related regulation and aid the investment industry to tackle climate change. Despite the positive approaches of both initiatives, they both share one thing in common – a complete disregard for tackling e-waste.

A closer look at e-waste

E-waste is the fastest growing waste stream in the EU and less than 40% of mobiles, laptops and other corporate electronics are recycled.

In 2019, more than 53 million metric tonnes of e-waste was produced. Current approaches to device and IT asset lifecycles are simply unsustainable, and Europe continues to dispose of electronic waste in some of the poorest areas in the world, where parts are stripped down and burned at the detriment of people's health. Not only that, but these e-waste graveyards represent a significant security risk, with the data on inappropriately sanitised equipment being intercepted by hacker groups.

The current challenge is indicative of a lack of proper awareness on dealing with end-of-life IT equipment. According to research, over a third of enterprises physically destroy their used assets because they believe it is 'better for the environment', however, this only contributes further to landfill and the rising tide of e-waste. More must be done to engage with the circular economy, and organisations will realise that in doing so, they will unlock the latent value in used devices and IT equipment while strengthening their security posture.

'E-waste is the fastest growing waste stream in the EU and less than 40% of mobiles, laptops and other corporate electronics are recycled.'

There's life in the old dog yet

More often than not, computers, servers, smartphones and laptops end up as e-waste because they are partially or completely physically destroyed. While proper physical destruction with an audit trail and certification is a valid method of data sanitisation in line with regulatory standards, it is almost entirely unnecessary. If, for example, the hard drive is removed from a system and is shredded, then the remaining system is most likely not reused, making it a far greater risk that the remaining components end up in landfill.

IT equipment such as hard drives can have their data erased with software – something that can now be done entirely remotely – meaning that the device and/or its parts can be reprocessed and resold without fear that residual data will be intercepted. For solid-state drives (SSDs), for example, secure SSD erasure uses technology and processes that erases the data all the way down into the over-provisioned cells, and provides verification that complete sanitisation has occurred to the highest 'Purge' security level. Drives are processed still in systems so there is no need to spend resources on removing drives for external processing. Having achieved erasure, the drives and entire systems can then be refurbished for reuse or resale. Only damaged or broken drives not responding to software sanitisation mandate physical destruction.

Securing a sustainable future

Essentially, organisations both public and private must explore methods of extending these device lifecycles and engaging with the circular economy. This would see devices be repurposed, repaired, resold or recycled, following data erasure. That refurbished hardware or equipment could then be reused internally, provide affordable options for second-hand purchasers, or even be safely donated to organisations and communities in need. And if their IT equipment becomes truly non-functional, they should send it to an environmentally friendly recycler or ITAD for certified data sanitisation via physical destruction in a way that enables proper recycling without risking the data being breached.

Many public institutions, as well as private businesses, are unfortunately still running destruction processes for hard drives instead of complete system re-use processes. These are legacy procedures that are outdated and must change. The public sector, in particular, should lead the way here and work to minimise unnecessary physical destruction, reducing e-waste, while also prioritising data sanitisation methods that guarantee the irreversible removal of sensitive data from a device.

To do so, sustainable practices should be outlined by a clearly defined e-waste policy or CSR policy that includes e-waste. It's crucial to call out end-of-life data management within this policy, clarifying roles and responsibilities, and highlighting how the organisation is helping to combat e-waste. This will not only help drive sustainability forward but also foster a culture of improved cyber-hygiene. Where possible, the circular economy should always be considered too, to ultimately change attitudes to e-waste and reduce landfill contributions.

5 November 2021

Can the world feed 8 billion people sustainably?

Food production is a major factor in the climate crisis and still millions of people go hungry. Here are some potential solutions.

By Weronika Strzyżyńska

More than enough food is produced to feed all of the 8 billion people currently alive on the planet, yet after a decade of steady decline hunger is back on the rise, affecting 10% of the global population. According to the World Food Programme, ripple effects of the Covid-19 pandemic and the war in Ukraine have contributed to one of the worst food crises in decades, with acute food insecurity affecting 200 million more people globally than in 2019 due to rising costs of food, fuel and fertiliser.

But there are bigger problems on the horizon. As the global population passes 8 billion and is predicted to reach 10 billion by 2050, farmers, governments and scientists face the challenge of increasing food production without exacerbating environmental degradation and the climate crisis, which itself contributes to food insecurity in the global south.

The United Nations projects that food production from plants and animals will need to increase 70% by 2050, compared with 2009, to meet increasing food demand. But food production is already responsible for nearly a third of carbon emissions as well as 90% of deforestation around the world.

'We use half of the world's vegetative land for agriculture,' says Tim Searchinger, a researcher at Princeton University. 'That's enormously bad for the environment. We can't solve the current problem by moving to more intensive agriculture because that requires more land.

'We need to find a way to decrease our input [land] while increasing our food production.'

But there is no magic bullet to achieve this goal. Instead, an overhaul at every step of the food production chain, from the moment the seeds are planted in the soil to the point where the food reaches our dinner tables, will be necessary.

Shifting towards regenerative agriculture

For most of human history, agriculture consisted of sustenance farming – people cultivated crops and livestock to feed their households rather than to sell them for profit. This began to shift after the Industrial Revolution and emergence of market capitalism, which also saw the rise of plantation farming made possible by colonisation of overseas land and slave labour.

Industrial farming not only increased the scale on which crops were cultivated but changed the techniques used by farmers. Instead of rotating the crops that were grown on a field each year, entire plantations would be dedicated to a single crop. This monocultural approach coupled with intensive modes of farming led to destruction of local biodiversity and land degradation – within years fields would cease to produce crops.

'The current agricultural paradigm is that land is cheap and infinite'

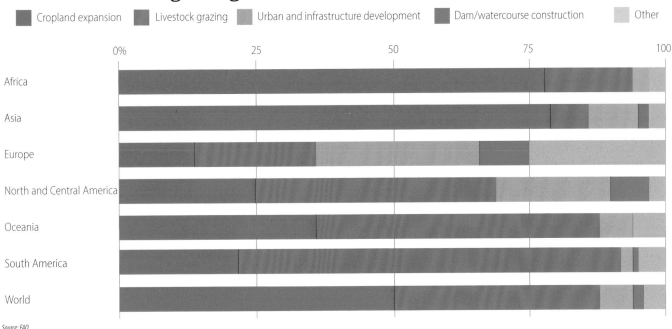

Deforestation is driven by agriculture in Africa and Asia and by grazing in Oceania and the Americas

■ Cropland expansion ■ Livestock grazing ■ Urban and infrastructure development ■ Dam/watercourse construction ■ Other

Source: FAO

Plantations of the 18th and 19th centuries were a 'get rich quick scheme' rather than a stable long-term investment, says Frank Uekötter, a professor of environmental humanities at the University of Birmingham. Plantation owners would extract maximum profits in a short period of time from their land. Once a field became unusable they would simply move on to new land. 'Up to the end of the 19th century, wide swaths of our planet were still not claimed by global modernity,' says Uekötter.

But today, while we are quickly running out of vegetative land, this colonial-era mindset persists. 'The current agricultural paradigm is that land is cheap and infinite,' says Crystal Davis from the World Resources Institute. 'Most farmers just cut down more trees, when new land is needed.'

'But to meet our ecological goals, we need to halt the conversion of natural ecosystems into farmland,' Davis says. 'We can achieve this in part by restoring degraded land back to its ecological integrity and productivity.'

Land restoration does not have to mean bringing it back to its original, pre-agricultural, state. 'There's a hybrid solution where we are bringing trees and other natural elements back to the landscape while also integrating production systems,' Davis says. 'Systems that are integrated with trees and other plants often are more sustainable and more productive over the long term.'

Davis points to Initiative 20 x 20, which has seen 18 South American and Caribbean countries, including Argentina and Brazil, commit to restoring 50m hectares of land by 2030. The initiative includes a number of projects aimed at introducing agroforestry practices to cocoa and coffee farms in Colombia and Nicaragua, where farmers are encouraged to grow crops while introducing more trees to their land.

Cutting food miles by growing crops locally

Transportation is a key, if often overlooked, part of the food production chain. Crops are transported from farms to processing plants before the food products arrive in shops. Packaging and transportation of food is responsible for 11% of all food industry greenhouse gas emissions. The emissions are not only caused by petrol used by trucks, which transport food across countries and continents, but also the refrigeration systems necessary to keep the produce fresh on its journey.

Freight transport contributes significantly to the carbon footprint of fruit and vegetables, releasing almost twice as much greenhouse gases as the process of growing the crops. This means that to reduce the environmental impact of food production, a shift towards plant-based diets in wealthier countries has to be coupled with more locally grown produce.

'In the UK, roughly half the food comes from this country and half comes from other places around the world – that has a large carbon footprint,' says Madeleine Pullman, a professor of sustainability and innovation at the University of Sussex. A solution for countries like the UK, Pullman says, is to increase the diversity of food that is produced domestically by allocating subsidies to farmers to grow a wider range of fruits and vegetables.

But in lower-income countries with hot climates, transportation poses a different challenge, as refrigeration of produce during transit is costly, meaning that much of the food is spoiled or incubates bacteria before it reaches customers.

'It's not always appropriate to move a western-style cooling system into a place in, for example, Africa,' Pullman says, pointing to Rwanda which introduced a national cooling strategy in 2018. Among other solutions, the plan includes subsidies for farmers to buy more efficient cooling equipment and trialling solar-powered cooling facilities.

'In Europe, we pay a lot of money to have food that has been moved and kept refrigerated, but when the vast majority are living in poverty, they cannot afford that,' Pullman says.

Abdulraheem Mukhtar Iderawumi, researcher at Oyo State College of Education in Nigeria, says that improving rural infrastructure such as roads and bridges would make transportation of harvested goods more efficient for smallholder farmers. He also suggests increasing farmers' access to trucks specially designed for transporting food as well as sharing information on best practice. 'Transportation should be done early in the morning or late in the evening,' he says. 'That is the time period when humidity is less of a risk to produce.'

Eating less meat

Shifting dietary habits is one of the most necessary solutions to the climate crisis, but it is also one of the most

Breakdown of greenhouse gas emissions from food production

24% Land use

27% Crop production

31% Livestock and fish farms

4% Processing

5% Packaging

6% Transport

3% Retail

Source: Our World in Data/Reducing food's environmental impacts through producers and consumers, Poore and Nemecek (2018)

controversial and difficult to introduce. More than half of all carbon emissions from the food industry are due to production of meat and animal-based products. Beef production emits more than twice as much CO2 a kilo of food as other types of meat produce, and 20 to 200 times more than plant products such as cane sugar or citruses.

Currently, 77% of agricultural land worldwide is used for the production of animal-based products. This includes a third of all cropland, as grains and crops are grown to produce animal feed and biofuel rather than for human consumption.

'Any global problem that you have, food is implicated in it,' says Tara Garnett, a researcher at the University of Oxford. 'On the one hand there are environmental problems associated with food, on the other there are health-related problems such as malnutrition, obesity and diabetes.'

> 'The approach to promoting insect diets is to disguise them in such a way that you wouldn't recognise a live insect.'

Garnett worked on the EAT-Lancet Commission, which in 2019 published its report on *Planetary Health Diet*. 'The idea was to figure out if there is a way of feeding everyone in a nourishing way on this planet, in ways that don't cause environmental harm,' Garnett says.

The diet can be best described as 'flexitarian'. Meat and dairy constitute important parts of the diet but in significantly smaller proportions than whole grains, fruits, vegetables, nuts and legumes. The diet recommends eating no more than 98 grams of red meat, 203 grams of poultry and 196 grams of fish a week.

'If you were to adhere to that diet, it would mean a massive reduction in meat, and to a lesser extent, dairy consumption in the global north, but it would actually give rise to more animal product consumption in many low-income countries,' Garnett says.

However, implementing lifestyle changes among a whole population is difficult.

'[The report] caused a lot of controversy, some saw it as a kind of a "vegan agenda",' Garnett says. 'There hasn't been a country that has adopted the diet as its national dietary guideline.'

She adds: 'Meat reduction is a very contested and value-laden idea that is perhaps kind of more personal than, for example, switching your boiler.' But she argues that changing dietary habits cannot be achieved by focusing on individuals. 'All the drivers, all the incentives and the disincentives, are currently working against the ability of people to eat and behave differently,' she says. 'Stop blaming the individual is one point I would make. There is a much greater role for government leadership and the food industry to play.'

Bamidele Raheem, a researcher at the University of Lapland, believes that dramatic changes in dietary habits might require generational change.

'Younger generations seem to be more curious about alternatives,' he says of his research on entomophagy, the technical term for eating insects.

Insects, which are commonly eaten in parts of Africa, Asia and South America, can be a more sustainable alternative to meat protein. 'They are much easier to rear than cattle. They can be produced in a much smaller space at a much higher rate and can be fed on food waste,' Raheem says. 'They are also richer in essential nutrients, such as iron, calcium and zinc.'

But westerners, who are the biggest consumers of red meat, face substantial mental barriers to enriching their diets with insects. 'This is where the mindset comes in,' Raheem says.

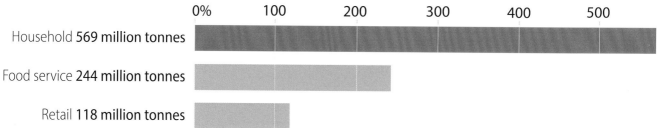

Households wasted an estimated 569m tonnes of food globally in 2019

	0%	100	200	300	400	500
Household 569 million tonnes						
Food service 244 million tonnes						
Retail 118 million tonnes						

Source: UNEP Food Waste Index report 2021

'The approach to promoting insect diets is to disguise them in such a way that you wouldn't recognise a live insect. For example, powdered crickets can be mixed with bread flour to make baked goods.'

The European Union has recently approved house crickets, yellow mealworms and grasshopper to be sold in frozen, dried and powdered forms. Raheem thinks we could see baked goods made using the insect ingredients commonly sold in Europe within the next five years.

In 2019, only 9 million people across the EU were estimated to be consuming insect-based products, but the International Platform of Insects for Food and Feed forecasts this number could reach 390 million by 2030.

While meat consumption in the west seems to be slowing down, and self-reported consumption of meat in the UK fell by 17% between 2008 and 2018, researchers credit this with raising awareness of the ecological downsides of meat rather than specific initiatives.

Reducing food waste and loss

An estimated third of all produced food is never eaten, according to the UN, with 14% of food lost between harvest and retail, and another 17% thrown out by shops, restaurants and consumers.

Food 'loss' rather than 'waste' describes the food that never reaches consumers. This problem is more prevalent in low-income countries where farmers cannot afford secure storage facilities and refrigeration. 'When there are no proper storage facilities the crops can be destroyed by the rain,' says Abhishek Chaudhary, a researcher at the Indian Institute of Technology Kanpur.

> 'Consumers in rich countries need to be made aware of how much food they're wasting.'

In Kenya, for example, smallholder farmers who produce more than 90% of the country's fruit and vegetables lose half of their harvest before they are able to sell it. 'Better storage facilities will require a lot of technology transfer from richer countries to poorer ones and a holistic approach,' says Chaudhary.

An example of this could be a ColdHubs initiative in Nigeria, which allows farmers access to pay-as-you-go solar-powered cold rooms. The company currently operates 54 refrigeration units in 22 states across the country.

In the global north, however, the problem of food waste – that is, food which is never eaten after it is sold – is more prevalent than food loss. According to a UN report, 931m tonnes of food is thrown away every year, with most waste occurring in households.

'The consumers in rich countries need to be made aware of how much food they're wasting,' says Chaudhary. 'Big food companies also have responsibility. If they can design and label the product smartly, then the consumers who are buying it will waste less food. For example, they can make the package size smaller. If you have a packet of chips, for example, but people don't usually eat all of it, then smaller packet is better.'

Digital data collection can also be used by shops, supermarkets and restaurants. 'By using smart data, retailers can see which things consumers are buying and adjust their inventory,' says Chaudhary. 'Individual households can also keep a food diary to see which products they end up throwing away.'

15 November 2022

Key Facts

- The United Nations projects that food production from plants and animals will need to increase 70% by 2050.

- Packaging and transportation of food is responsible for 11% of all food industry greenhouse gas emissions.

- 77% of agricultural land worldwide is used for the production of animal-based products.

- An estimated third of all produced food is never eaten, according to the UN with 14% of food lost between harvest and retail and another 17% thrown out by shops, restaurants and consumers.

- According to a UN report, in the global north, 931m tonnes of food is thrown away very year, with most waste occurring in households.

Food for the future: understanding the challenges and options for sustainable food production in England using recirculating aquaculture systems

By Keith Jeffrey

With increasing concerns around food security in the face of growing populations, climate change and recent global events, such as the Ukraine invasion, the UK government is turning their attention to new and innovative ways of meeting future food demands and tackling rising prices.

One method of focus is the use of Recirculating Aquaculture Systems (RAS). Unlike traditional aquaculture where fish are grown outdoors, RAS is a modern and highly technical form of aquaculture, which uses indoor tank systems to grow fish in a more controlled environment. By using a closed-loop system that recirculates water, RAS minimises the risks associated with conventional fish farming practices such as escapees, pollution via effluents, predation, biosecurity, water usage, and habitat preservation. While RAS development in England has been slow in the last three decades, it has gained international recognition due to the increasing scale and technological improvements of the systems.

The role of RAS in the English Aquaculture Plan

The UK government was interested in the potential for RAS to help achieve the targets set out in the English Aquaculture Plan (34,608 tonnes of aquatic food production per year by 2040) whilst also achieving the sustainability objective within the Fisheries Act 2020. To understand the feasibility of RAS production in England, as well as the potential challenges, scientists at the Centre for Environment, Fisheries, and Aquaculture Science (Cefas) conducted an extensive literature review and stakeholder survey. This comprehensive study engaged various stakeholders, including RAS funders, operators, regulators, consultants, academia, Non-Governmental Organisations (NGOs), and trade organisations.

Environmental, regulatory and economic challenges

The survey results revealed that despite increasing global interest, RAS in England are currently limited to small-scale aquaculture farms, hatcheries and holding systems with a few larger scale systems in early developmental stages. However, global growth in RAS continues apace with systems being built capable of producing 5 to 120 thousand tonnes per year.

Within RAS literature a common finding was a concern about energy costs and the carbon footprint of the systems. However, efforts are underway within the industry to reduce the energy cost per kilo of production through design and operational efficiencies, use of renewable energy sources, reuse of waste, and locating production closer to markets to improve fuel efficiency.

Findings from the survey highlighted that despite the advantages of RAS, operators in England face challenges related to site selection for larger production units, as well as the time and costs associated with obtaining licences and permissions. The growth of RAS systems is thought to be most likely on brownfield sites and terrestrial farms where infrastructure already exists, and simplified licensing and permitting systems are in place.

Another challenge identified in the study is the need for specialised skills for a RAS workforce. Traditional aquaculture skills do not easily transfer to RAS, highlighting the importance of universities and other educational institutions in developing new RAS designs whilst providing skilled graduates to support the sector's growth.

Operational/production costs are another barrier to the widespread adoption of RAS within England, with industry stakeholders highlighting the high initial costs and regulatory barriers as the main obstacles limiting the sector's growth. Given that initial costs are higher than for conventional aquaculture, RAS systems are economically vulnerable due to long break-even periods, which are further extended when unforeseen events such as power failures or disease outbreaks occur. To overcome these challenges, the industry has called for increased financial incentives for research and development stages and support through the early stages of start-up.

The path to sustainable aquatic food production

RAS offer a promising solution to some of the challenges faced by the aquatic food system and the need for sustainable seafood production. While RAS development in England has been limited, international investments and growth, along with emerging technologies, demonstrate the potential for RAS to contribute significantly to global aquatic protein supply.

Overcoming regulatory barriers, addressing the carbon footprint, reducing operating and capital costs and incentivising research and development are crucial steps towards realising the full potential of RAS. If England embraces RAS, the review and survey suggest that England could enhance its aquatic food production figures, and achieve the targets within the English Aquaculture plan whilst using a minimal spatial footprint in a way that removes many traditional impacts of aquaculture.

27 June 2023

Useful Websites

www.activesustainability.com

www.britishcouncil.org

www.currentnews.co.uk

ww.europarl.europa.eu

www.goodonyou.eco

www.gov.uk

www.independent.co.uk

www.openaccessgovernment.org

www.politico.eu

www.positive.news

www.sdgs.un.org

www.sgvoice.net

www.shoutoutuk.org

www.sustainability-success.com

www.theconversation.com

www.theguardian.com

www.un.org

www.weforum.org

www.ypte.org.uk

Further Reading
Pages 18-19:
https://www.europarl.europa.eu/news/en/headlines/
economy/20151201STO05603/circular-economy-definition-importance-
and-benefits
Pages 20-21:
https://www.weforum.org/agenda/2021/12/fight-plastic-pollution-
innovations
https://www.sostenibilidad.com/vida-sostenible/desmontando-mitos-
reciclaje/

Glossary

Biodegradable waste

Materials that can be completely broken down naturally (e.g. by bacteria) in a reasonable amount of time. This includes organic materials such as food waste, paper waste and manure, which can be composted, as opposed to items such as plastic bottles that would take thousands of years to break down naturally.

Eco-friendly

Policies, procedures, laws, goods or services that have a minimal or reduced impact on the environment.

E-waste

Electronic waste; discarded electrical items such as mobile phones and computers. There are strict EU regulations in place to ensure that e-waste is safely recycled or disposed of: however, the shipping of e-waste to developing countries is becoming an increasingly common problem.

Ethical consumerism

Buying things that are produced ethically - typically, things which do not involve harm to or exploitation of humans, animals or the environment; and also by refusing to buy products or services not made under these principles.

Fair trade

A movement which advocates fair prices, improved working conditions and better trade terms for producers in developing countries. Exports from developing countries that have been certified Fairtrade – which include products such as coffee, tea, honey, cocoa, chocolate, sugar, cotton and bananas – carry the Fairtrade mark.

Fast fashion

Inexpensive, mass-produced clothing that is usually produced quickly to respond to current fashion trends. Often, it is only worn a few times before being thrown away.

Food waste

Around seven million tonnes of food is thrown away by households in the UK every year. Some of the waste is unavoidable, such as peelings or bones, but most of the food is edible. This is because there is often confusion over use-by and best-before dates. Also, many families buy more food than they actually need.

Global footprint

A person's global footprint refers to the impact that they have on the planet and the people around them, taking into account how much land and water each person needs to sustain their lifestyle.

Green energy

The same as renewable energy, which comes from natural resources rather than non-renewable sources. It's called 'green' due to the fact that the sources are environmentally friendly, sustainable and have zero emissions.

Greenwashing

'Greenwashing' occurs when organisations falsely promote or market themselves as having 'green', environmentally-friendly, practices.

Recycling

The process of turning waste into a new product. Recycling reduces the consumption of natural resources, saves energy and reduces the amount of waste sent to landfills.

Resource consumption

The use of the Earth's natural supplies, including fossil fuels, water, wood, metals, minerals and many others. Growing populations and increased standards of living have resulted in increased consumption of natural resources, which is having a negative effect on the environment.

Sustainability

Sustainability means living within the limits of the planet's resources to meet humanity's present-day needs without compromising those of future generations. Sustainable living should maintain a balanced and healthy environment.

Sustainable Development Goals (SDGs)

17 goals set out by the United Nations to protect the planet and ensure that people around the world can live with equality and in a healthy environment by 2030. The goals cover social, economic and environmental sustainability. 'End poverty in all its forms everywhere' is the number one SDG.

Sustainable diet

Sustainable diets have a low environmental impact – this includes the impact of food production and consumption on our planet's resources.

Waste

Anything that is no longer of use and thrown away. Each year the UK generates approximately 290 million tonnes of waste, which has a damaging effect on the environment.

Wishcycling

The act of putting something in the recycling bin in the hope it can be recycled even though it might be unsuitable.

Index